AF894145

Responses to Territorial Revision

Historical Lessons

ANDREW STRAVERS, EMILY ELLINGER, GREGORY WEIDER FAUERBACH,
CLARE PORTER

Prepared for the Department of the Air Force
Approved for public release; distribution is unlimited.

RAND PROJECT AIR FORCE

For more information on this publication, visit **www.rand.org/t/RRA3140-2**.

Published by the RAND Corporation, Santa Monica, Calif.

Library of Congress Cataloging-in-Publication Data is available for this publication.

ISBN: 978-1-9774-1517-2

Cover: Imagedepotpro/Getty Images.

About This Report

In this report, the authors evaluate the factors that drive regional and third-party responses to successful territorial revision. They found that third-party states take into account a variety of factors involved in the territorial revision when determining their alignment decisions in its aftermath. These factors included fears of successive revision attempts, continued credible involvement from another outside power, legitimacy according to international and regional organizations, economic links between relevant parties, and the treatment of the population after a territorial takeover.

The research reported here was commissioned by the Department of the Air Force and conducted within the Strategy and Doctrine Program of RAND Project AIR FORCE as part of a fiscal year 2024 project, "The Day After . . . Taiwan."

RAND Project AIR FORCE

RAND Project AIR FORCE (PAF), a division of RAND, is the Department of the Air Force's (DAF's) federally funded research and development center for studies and analyses, supporting both the United States Air Force and the United States Space Force. PAF provides the DAF with independent analyses of policy alternatives affecting the development, employment, combat readiness, and support of current and future air, space, and cyber forces. Research is conducted in four programs: Strategy and Doctrine; Force Modernization and Employment; Resource Management; and Workforce, Development, and Health. The research reported here was prepared under contract FA7014-22-D-0001.

Additional information about PAF is available on our website:

www.rand.org/paf/

This report documents work originally shared with the DAF on October 15, 2024. The draft report, dated February 2025, was reviewed by formal peer reviewers and DAF subject-matter experts.

The views expressed in this report are those of the authors and do not reflect the official policy or position of the Department of Defense or the U.S. Government. Review of this material does not imply Department of Defense endorsement of factual accuracy or opinion.

Acknowledgments

We gratefully acknowledge the input of the sponsor's office, notably Col Richard Ganske, in developing the project's findings. We appreciate the support of the leadership of the Strategy and Doctrine Program, Raphael Cohen and Alexandra Evans. We are grateful for comments from our two peer reviewers, Scott Harold and Paul Heer.

Summary

Issue

This report asks, "What factors drive third-party state decisions after territorial revision?" It uses quantitative and historical analysis to identify some of these broad factors to understand what the likely third-party calculations will be after potential future territorial revisions.

Approach

To assess what drives third-party state decisionmaking, we first conducted a quantitative analysis to determine the frequency of territorial takeover attempts and their success. We then selected and reviewed five historical examples of territorial revisionism to derive lessons about possible geopolitical reactions.

Key Findings

- *Territorial revision is a rare event, and successful conquest is even rarer.* Since World War II, there have only been 70 conquest attempts, of which 29 resulted in the conqueror holding the conquered territory ten years later.
- *Revision attempts tend to raise regional state threat perception. Spillover effects*—in this case, the possibility of continued territorial seizure by the successful conqueror—drove behavior to guard against further attempts at territorial revision.
- *States consider a conqueror's adherence to international norms when deciding their response to a territorial revision.* Third-party countries take a revisionist's failure to abide by these norms—especially in its treatment of "conquered" populations—as an indicator of the continuing threat a conqueror poses and its possibly broader territorial ambitions.
- *Strong international institutions facilitate reactions to revisionism.* Institutions serve as fora to discuss and decide on collective policy responses to territorial-revision attempts. They also provide potential conquerors incentives to follow the international order's norms, rules, and procedures.
- *Economic links play an important role both during and after revisionism attempts.* States' economic considerations play a vital role in determining their alignment decisions. Third-party countries' attempts to both avoid economic pain and seek economic benefits were found in the case studies.

Implications and Recommendations for the Department of the Air Force

The historical analysis can inform the United States' broad geopolitical strategy, as well as strategic assessments of potential contingencies after revisionism attempts around the world. The findings also have several more-specific implications for the Department of the Air Force (DAF):

1. Revisionism attempts anywhere in the world would prompt a complicated set of calculations from regional actors. A key role for the DAF is to provide continued involvement from U.S. forces by having necessary regional access and offering credible plans and strategies to protect regional states from further revisionism attempts.
2. Concurrently, there is a need to invest in a better understanding of partner and ally interests in response to a successful takeover attempt. This is to determine which activities allies and partners view as signaling the credibility of the United States' commitment to their security interests following territorial takeover. These activities include facilitating partner military modernization, arms acquisition, and tightened security partnerships with states whose threat perceptions are raised in the aftermath of a takeover.
3. Advanced and active engagement in international institutions can facilitate coalition-building in the event of a takeover attempt. Conversely, failure to adhere to established norms can weaken institutions. Efforts to enforce and frame norms can drive threat perceptions and resistance to revisionism and harsh treatment of acquired populations. Such activities as International Military Education and Training and hosting liaison officers can assist in instilling such norms across partnerships.

Contents

Figure and Tables

Figure

Tables

Chapter 1

Territorial Revisionism: Quantitative Evidence and Qualitative Methodology

What factors do states consider when determining their alignment in the aftermath of a territorial-revision attempt? This report considers five cases that span peaceful cession, resistance with little international support, and resistance with significant international support. Each case occupies a different place on this spectrum, with the case of Hong Kong featuring a fully peaceful cession and the Soviet invasion of Czechoslovakia seeing some very limited local resistance. The Japanese invasion of Manchuria saw full-scale local resistance with no international support. At first, the Iraqi invasion of Kuwait saw significant Kuwaiti resistance with no immediate international support, followed by a full-scale coalition response. Last, in the case of the Crimean War, the conflict was internationalized from the very beginning.

Figure 1.1. Spectrum of Conflict and International Involvement with Case Selection

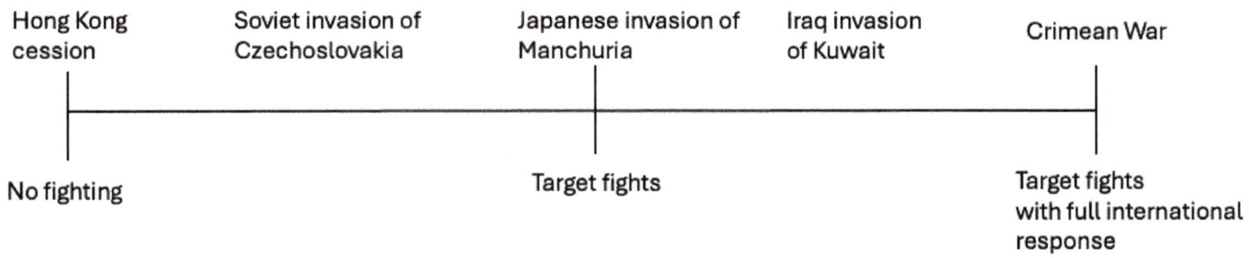

Our study used quantitative and qualitative analysis to evaluate this question. We found that there are several key factors that states use to determine their responses. If the successful seizure of territory positions a revisionist state to more easily take over additional territory or exert additional leverage over neighboring states, third parties may see a heightened threat perception that can drive alignment behavior. States can then choose a variety of options to mitigate this additional threat, such as aligning more closely with other states that oppose additional revisionism.

Alignment behavior, defined as the political actions that states take toward one another, responds to possible increased threat perception, but it is mediated by additional factors. First, third-party states must see continued credible involvement from an outside great power to believe that they have options to resist the *revisionist state*, defined as a state taking actions to upend a territorial status quo.[1] Great powers can thus galvanize counter-revisionist coalitions, in part by assuring third parties that their security interests will be protected. Second, international institutions play a key role in facilitating collective action among smaller third-party states to ensure that those states are protected against isolation and punishment by the revisionist state. Third, economic links play a powerful mediating role in considerations of alignment by either constraining or freeing third-party states. When their economic livelihoods are connected to the revisionist state, third-party states are less able to actively align against a revisionist state because the revisionist can inflict punishment on third-party states without needing to resort to military action. Table 1.1 outlines these key factors and the cases to which they apply.

Table 1.1. Key Factors in Determining Regional and International Alignment in Historical Cases

Historical Case	Fear of Successive Takeovers	Continued, Credible Involvement from Great Powers (or Lack Thereof)	International and Regional Organization Legitimacy	Economic Links	Harsh Treatment of Target Population
UK cession of Hong Kong 1997	✓	✓	✓	✓	✓
USSR invasion of Czechoslovakia 1968	✓	✓		✓	✓
Japan invasion of Manchuria 1931	✓	✓	✓	✓	✓
Iraq invasion of Kuwait 1990	✓	✓	✓	✓	✓
Crimean War 1853–1856	✓	✓		✓	✓

NOTE: UK = United Kingdom; USSR = Union of Soviet Socialist Republics.

Although this work is not an exhaustive study of all factors that third-party states consider—it excludes, for example, their relationships to each other and the thresholds that tip a state in one direction or another—it does identify consistent factors that emerged in historically and regionally disparate cases. Therefore, they are likely to be important considerations for alignment decisions in the aftermath of most attempts at territorial revision.

This project has important implications for U.S. strategy in many places. China has long claimed Taiwan as part of its territory, and fears of a successful takeover drive security behavior in Japan and other regional partners. Fears of further Russian incursions into Europe have driven the behavior of many European allies, with the North Atlantic Treaty Organization (NATO) standing as a key organizer of efforts to respond to the Russian invasion of Ukraine, and U.S. natural gas providing a

[1] In this report, *great power* is used to refer to a country capable of significantly influencing events outside its own region.

credible alternative for countries that rely on Russia. Similar considerations are also present in the Middle East, with persistent Iranian attempts to change the status quo in the region.

This chapter begins by using quantitative data to establish a baseline understanding of territorial revisionism. We then detail how we selected historical cases, which are then explored in subsequent chapters.

Quantitative Analysis of Territorial-Revision Attempts

Although the broader findings of our research relate to the idea of territorial revisionism, which includes territorial changes through conflict and by peaceful means, our quantitative work focused on the concept of territorial conquest. This was done for two primary reasons. First, most future scenarios do not consider territorial revision without conflict in areas of strategic importance to the United States, especially when considering main U.S. competitors and the places they threaten.[2] Second, generally, there is more documentation on and analysis of conflictual incidents of revisionism, which offer a better understanding of the decisions and considerations of third-party states.

To conduct a high-level quantitative examination of conflictual revisionism attempts, we drew on Altman's Modern Conquest dataset (version 2.2) to baseline the frequency of territorial conquest and identify the universe of cases.[3] This analysis produced the following two major findings:

1. Territorial conquest attempts have an about even chance of succeeding in the short term (49 percent) and a slightly lower chance of both taking and holding the conquered territory for at least ten years (44 percent).
2. Third parties rarely fire a shot to defend targets of conquest attempts, even when that victim is part of a defense pact. In the rare cases in which a third party *does* directly intervene on behalf of a victim, the victim is somewhat likelier to hold onto its territory.

Altman's dataset defines a conquest attempt as follows:

> A conquest attempt occurs when one state uses its military to seize disputed territory without permission and with the intention to assume lasting control of that territory.[4]

The appendix contains a list of all cases used for this analysis. To judge whether a case qualifies as a *conquest attempt,* the dataset relies on two indicators:

- The articulation of a sovereignty claim to the territory, and . . .
- the establishment of a fixed position in the disputed territory where before there was none (either because it was empty or because the defender held that ground).[5]

---

[2] We still consider revision without conflict in the case studies discussed in this report to highlight factors that third-party states considered in the aftermath of a nonconflictual change to the territorial status quo to determine whether these factors are substantively different from factors after conflict.

[3] Altman's Modern Conquest dataset (version 2.2) is available at Dan Altman, "Replication Data for: 'The Evolution of Territorial Conquest After 1945 and the Limits of the Norm of Territorial Integrity,'" Harvard Dataverse, Vol. 1, 2020.

[4] Altman, 2020.

[5] Dan Altman, "Codebook for the Modern Conquest Dataset Version 2.2," January 10, 2021.

A single conflict can contain multiple conquest attempts (such as World War II). Each case in the dataset has a *perpetrator*, defined as "the state that seizes the territory," and a *victim*, defined as "the state from whom territory is taken."[6] These labels are imperfect—they should not be seen as normative judgments of whom a territory truly belongs to, and they are not meant to label one state as responsible for the conflict in which a conquest attempt occurs.

In coordination with the project sponsor, we focused on conquest attempts from 1946 to 2017.[7] We exclude cases that the dataset labels *retaliatory*, which is a category comprising cases, such as the USSR's retaking of its own territory that had been seized by Germany earlier in World War II or the United States' retaking of the Philippines during World War II.[8]

Table 1.1 breaks down the conquest attempt dataset. Most importantly, territorial conquest attempts are rare events: There were only 70 from 1946 to 2017.[9] Most conquest attempts involved combat, although a significant minority (31 percent) did not see troops killed. Territorial conquest attempts are also reasonably successful: Perpetrators gain and hold territory in nearly half of all attempts. Moreover, as seen in both Tables 1.2 and 1.3, a majority of victim countries were members of a defense pact at the time of the conquest attempt.

However, it is likely that perpetrators targeted alliances in which commitments were weak: Third parties defended the victim in only 7.3 percent of cases and abandoned most allies that faced conquest attempts. Indeed, third parties (with or without an alliance commitment) intervened in only 8.5 percent of conquest attempts. That said, as seen in Table 1.3, intervention significantly increases the chance that a victim retains its territory.

---

[6] Altman, 2021.

[7] A single conflict can contain multiple conquest attempts.

[8] The Modern Conquest dataset's codebook gives the following definition of *retaliatory*:

> This marks all conquest attempts that are not initial conquest attempts (It may be clearer to think of this variable as if it were named "NON-INITIAL"). All retaliatory conquest attempts occur within the same conflict (militarized dispute, crisis, or war) as the events to which they are retaliating/following. Retaliatory conquest attempts include the following: 1) Cases where the victim retakes land that the perpetrator just seized; 2) Cases where the victim takes more or different land than the perpetrator just seized; 3) Cases where the victim retakes the seized territory and then continues on to attempt to remove the regime of the perpetrator; 4) Cases where the victim of an entire-state conquest attempt that failed to take complete control takes back seized territory in a part-of-state retaliatory conquest attempt; 5) Cases where the conquest attempt begins during a war that is already underway for other reasons. (Altman, 2021)

[9] We chose to start our *statistical* analysis in 1946 for two key reasons. First, it provides a period with complete data coverage required to perform the analysis (1919–2017). Second, our sponsor expressed a preference for non–World War II cases in the statistical analysis, as well as cases occurring after the emergence of nuclear weapons. For our study, we performed a statistical analysis that included World War II cases to understand the nature of any bias caused by their inclusion, and the results were largely the same as those presented here. The qualitative cases did not need to meet these criteria, so two cases are outside the bounds of the post–World War II time frame: the Japanese invasion of Manchuria and Crimean War.

Table 1.2. Territorial Conquest Attempts (1946–2017)

Features	*N* = 70
Violence (≥1 battle death)	
Yes	48 (69%)
No	22 (31%)
Third party–defended victim[a]	
Yes	6 (8.8%)
No	62 (91%)
Unknown	2
Victim in defense pact	
Yes	41 (59%)
No	29 (41%)
Victim held territory (immediately after)	
Yes	36 (51%)
No	34 (49%)
Victim held territory (ten years later)[b]	
Yes	37 (56%)
No	29 (44%)
Unknown	4

SOURCE: Authors' calculations based on the Modern Conquest dataset (version 2.2) and the Alliance Treaty Obligations and Provisions dataset (Altman, 2020; Quality of Government Institute, "Data Source: Alliance Treaty Obligations and Provisions Project [ATOP]," webpage, Quality of Government Data Finder, undated).

NOTE: The two unknown cases for the "third party–defended victim" variable are cases in which we disagreed with the Modern Conquest dataset's original coding but did not want to violate the dataset's original coding rules. For example, Iraq's 1990 invasion of Kuwait was originally coded as having no third-party defense because the United States–led coalition intervened only after Iraq had fully occupied Kuwait.

[a] *Third party–defended victim* means that a third party fired at least one shot in defense of the victim. See Altman, 2021.

[b] The four unknown cases for *victim held territory (ten years later)* correspond to cases in which either ten years have not elapsed since the initial conquest or ten years have passed but the dataset has not yet been updated.

Table 1.3. Defensive Alliances and Third-Party Interventions

	Victim in Defense Pact	
Third Party–Defended Victim	**Yes**	**No**
Yes	3 (7.3%)	3 (10%)
No	37 (90%)	25 (86%)
Unknown	1 (2.4%)	1 (3.4%)
Total	41 (100%)	29 (100%)

SOURCE: Authors' calculations based on the Modern Conquest dataset (version 2.2) and the Alliance Treaty Obligations and Provisions dataset (Altman, 2020; Quality of Government Institute, undated).

Qualitative Case Selection

As seen in the data, territorial takeover attempts are rare events, and successful ones are even rarer: There have been only 29 such cases since 1946. Moreover, we are concerned about the multiple pathways through which territorial revision could occur into the future. This study posits the following three general paths:

1. an uncontested takeover that features no fighting
2. a scenario in which the target fights with no international support
3. a scenario in which the target resists a takeover attempt with significant international support.

With these three general scenarios in mind, we sought to find historical corollaries that could provide insight into the behavior of regional states and great powers in territorial-revision circumstances.[10]

The set of cases even fitting these criteria is limited. In addition, we sought to find historical corollaries that fit the following other criteria:

- The events were of significant regional and international interest.
- The events resulted in significant changes to regional or international relationships.
- The target of the takeover attempt had been independently governed on an established territory with significant international recognition.
- The events involve an external military crossing preexisting territorial boundaries.
- The events result in an alteration of security issues for regional states.
- The events result in an alteration of international perceptions and linkages.

Other criteria were considered, which would provide for close corollaries on some measures *prior to* a takeover attempt. However, many of these variables were discarded because they did not provide useful insight into regional or international reactions. For instance, in many cases, there was simply a lack of international attention, there were few security implications for neighboring states, or the case

[10] To iterate the case selection point made in footnote 9, the qualitative case selection did not need to meet the criterion of being post–World War II.

was one in which regional or international actors did not have realistic options of intervention. Because the purpose of our study is to understand the specific factors that led states to shift alignment in the aftermath of a territorial-revision attempt, the study prioritized cases that saw significant attention in an effort to isolate the key determining factors of state behavior. Furthermore, we sought some regional and temporal variation to avoid the possibility that specific findings would be the result of a period or region of focus.

The resulting five cases that emerged from these considerations offer effective comparisons to future scenarios on any number of variables.[11] However, our purpose here is a limited one—to understand some key themes in how states have made decisions regarding their alignment in the aftermath of an attempted or realized territorial revision. The insights from these cases may be useful when contemplating the aftermath of similar events in the future.

The cases available and their key takeaways are summarized in the Table 1.4. Table 1.5 offers a summary of the historical cases of territorial revision that we analyze in this report. In the following chapters, we explore how spillover effects prompted regional state behavior in each of these cases. We also explore what measures revisionists take to mitigate balancing behavior, as well as how great powers have facilitated collective interventions and responses to successful territorial seizures.

Table 1.4. Third-Party Interventions and Conquest Outcomes

Victim Held Territory (Ten Years Later)	Third Party–Defended Victim		
	Yes	**No**	**Unknown**
Yes	4 (67%)	32 (52%)	1 (50%)
No	2 (33%)	26 (42%)	1 (50%)
Unknown	0 (0%)	4 (6.5%)	0 (0%)
Total	6 (100%)	62 (100%)	2 (100%)

SOURCE: Authors' calculations based on the Modern Conquest dataset (version 2.2) (Altman, 2020).

[11] We considered other cases for analysis, such as the Hungarian Revolution of 1956 and the People's Republic of China's (PRC's) conflict with Vietnam in 1979. However, we selected the five cases analyzed in this report in consultation with the sponsoring office to suit its interests, match team expertise, and provide some measure of regional variation within the constraints of project scope and budget.

Table 1.5. Historical Cases of Territorial Revision

Historical Case	Invading Country	Invaded Country	Supporting Invaded	Result
UK cession of Hong Kong 1997	China	Hong Kong	UK	Chinese takeover
Soviet invasion of Czechoslovakia 1968	Soviet Union	Czechoslovakia		Russian invasion
Japanese invasion of Manchuria 1931	Japan	China		Japanese takeover
Gulf War 1990–1991	Iraq	Kuwait	United States	Iraqi takeover, then expulsion
Crimean War 1853–1856	Russia	Ottoman Empire	France, UK	Russia takeover, then reversal

Chapter 2

British Cession of Hong Kong, 1997

Following the Opium Wars in the 19th century, the UK was granted the area surrounding Hong Kong. This expanded territory was extended in a 99-year lease in 1898, which granted the UK additional territory.[12] As the 99-year lease reached its termination, the UK and the PRC entered into negotiations on how the territory would be handed over to the PRC. Although British military defense or the retention of Hong Kong was never considered, this case highlights the importance of constraining spillover effects to legitimize territorial takeover. Beijing's takeover aligned with international norms in which colonial powers gave up their territories, as well as with British legal restrictions. However, the Tiananmen massacre of 1989 undermined Chinese assurances that it would abide by agreements made with the British and that it would respect Hong Kong's unique governance systems. Consequently, the UK scrambled to institute legal protections for Hong Kong citizens, although Beijing undid many of these protections over the next 25 years.

Examining the case of the Hong Kong cession to the PRC is critical to understanding a possible outcome of territorial revision without open conflict. Because Hong Kong was also not formally a separate state from the PRC, the reactions of regional states and the wider international system to the cession provide key insights into the outcome of territories with uncertain status. Nonconflict scenarios create a more gradual shift in power and place significant international weight on the diplomatic agreements made during the transition. In particular, a series of assurances given by the PRC to the UK and Hong Kong about the way in which Hong Kong would be governed after the cession—many of which were not kept by the PRC—had profound impacts on the way in which much of the international community would come to view the PRC's intentions and trustworthiness.[13] This negotiated territorial revision can provide context for the shifts in power and the long-lasting implications of such agreements.

Historical Context

The territory of Hong Kong was taken by the UK through several treaties over the course of the 19th century. In 1842, following the First Opium War, the Treaty of Nanking ceded the Hong Kong

[12] Convention Between the United Kingdom and China, Respecting an Extension of Hong Kong Territory, signed at Peking, June 9, 1898.

[13] Although there were disagreements over the nature and scope of exactly what was promised during the cession between the PRC and Hong Kongers, much of the international community sided with Hong Kong in this disagreement. As a result, the international community saw the PRC's actions as inconsistent with international human rights norms and as an example of a nondemocratic state's lack of trustworthiness in upholding its commitments.

islands to the UK "in perpetuity."[14] This territory was expanded by the Conventions of Peking in 1860, which permanently yielded the Kowloon Peninsula to the UK. In 1898, a new portion of land (dubbed *the New Territories*) was signed over to the British in a 99-year lease, which was set to expire on July 1, 1997.[15] This last section of territory accounted for 92 percent of the land of Hong Kong.[16]

Regional Alignment

Throughout the 20th century, the UK built up Hong Kong as a financial and business center and a gateway to the Asian markets. This was unchallenged by Chinese leadership and generally accepted by Hong Kong citizens and the international community.

Although the Hong Kong peninsula was temporarily occupied by Japan during World War II, Britain formally reestablished its control following the Japanese surrender.[17] Britain built Hong Kong from a largely empty peninsula into a strong financial center with substantial infrastructure, governmental bodies, education, health services, and sanitation systems.[18] Following the Korean War, the British administration of the colony shifted to a strategy of avoiding provoking China because British officials had become concerned that Britain would be unable to repel aggressive communist forces.[19] This was most notable in Britain's hesitancy to establish democratic procedures in Hong Kong. There was a belief among British officials that, if Hong Kong became fully democratic, it would move for full independence and, in reaction, China would invade.[20] The UK maintained this careful peace until the 1970s, when businesses that had grown rich in the century of British occupation began to look on the impending end of the Hong Kong lease with trepidation.[21]

Throughout this period, Hong Kong citizens had a unique relationship with the UK that distinguished their experience from that of people in a typical colonial situation. When Britain first took charge of Hong Kong, it lacked a settled population, infrastructure, and both a political and military presence from China.[22] As a result, most Hong Kong citizens moved to the city during British occupation, either to escape Chinese rule or to pursue economic opportunities. This led to a uniquely sanguine view of British colonial that was not often duplicated in other colonies.[23] The British overall policy for Hong Kong was to promote economic growth while maintaining social stability. However,

---

[14] Ming K. Chan, "Hong Kong: Colonial Legacy, Transformation, and Challenge," *Annals of the American Academy of Political and Social Science*, Vol. 547, No. 1, September 1996, p. 12.

[15] Chan, 1996.

[16] Colleen M. Granahan, "Note: The Prospects of a Successful Reunification of Hong Kong and China in Light of the Events at Tiananmen Square," *Boston University International Law Journal*, Vol. 10, 1992.

[17] Chan, 1996.

[18] Brian Hook, "British Views of the Legacy of the Colonial Administration of Hong Kong: A Preliminary Assessment," *China Quarterly*, Vol. 151, September 1997.

[19] Brian C. H. Fong, "Great Power Competition in Non-Sovereign Entities: US–China Tug-of-War over Hong Kong, 1950–2020," *Pacific Focus*, Vol. 37, No. 2, August 2022; Hook, 1997.

[20] Neville Maxwell, "Sino-British Confrontation over Hong Kong," *Economic and Political Weekly*, Vol. 30, No. 23, 1995.

[21] Fong, 2022.

[22] Lau Siu-Kai and Kuan Hsin-Chi, "Public Attitudes Toward Political Authorities and Colonial Legitimacy in Hong Kong," *Journal of Commonwealth and Comparative Politics*, Vol. 33, No. 1, 1995.

[23] Lau and Kuan, 1995.

British rule was nonrepresentative and rarely integrated with the local community, instead favoring the elite British establishment.[24] Despite their exclusion from the British colonial political system, Hong Kong citizens established a "social contract" with Britain, generally accepting the colonial system in exchange for economic opportunities and protection from communist Chinese control.[25]

The PRC did not militarily challenge Britain's system for ruling over Hong Kong because it saw Hong Kong as posing economic opportunity; nevertheless, PRC officials occasionally stated that they did not recognize the legitimacy of a treaty signed by China's previous imperial government.[26] Historical documents noted that, at the founding of the communist PRC, the leader Mao Zedong argued that Hong Kong was not strategically important enough to attempt to recover and the PRC would benefit more from using it to develop international and economic relationships.[27] This resulted in PRC troops stopping just outside the border of Hong Kong and, instead, focusing their interests inward throughout the 1950s and 1960s to solidify domestic communist rule.[28] Throughout this period, PRC leaders continued to treat Hong Kong as an important partner for economic development and advancement for China through the infusion of much-needed capital and innovative practices.[29] However, they also treated Hong Kong as a potential entry point for anticommunist activities from democratic nations.[30] Despite this hesitant peace, throughout the decades of the British lease, the PRC issued regular claims that the lease was invalid and the land rightfully belonged to China.[31]

Following prompting from British companies concerned about their business prospects in Hong Kong, the UK reached out in 1971 to begin negotiations for a new Hong Kong lease. Chinese officials stated that "no moves would be made against Hong Kong or Taiwan for a long time to come," but the following year China issued a formal letter to the UN in which it claimed that Hong Kong was an indivisible part of Chinese territory.[32]

This letter was an important turning point for the status of Hong Kong. Following World War II, the UN created a special committee that provided a legal basis to formally divest nations of their remaining colonies and included Hong Kong on the list of territories to which the Declaration of

---

[24] Political representation improved following the 1967 riots, leading Britain to create better social services, such as educational reforms, welfare programs, and labor laws (Chan, 1996).

[25] Michael Yahuda, "Hong Kong's Future: Sino-British Negotiations, Perceptions, Organization and Political Culture," *International Affairs*, Vol. 69, No. 2, April 1993, p. 246; Lau and Kuan, 1995.

[26] Huang Hua, letter dated 8 March 1972 from the permanent representative of China to the United Nations addressed to the Chairman of the Special Committee, United Nations (UN) document A/AC.109/396, March 9, 1972.

[27] Fong, 2022.

[28] Fong, 2022; Yahuda, 1993.

[29] In the 1960s, one-half of China's income was the result of selling water and food to Hong Kong. By the 1980s, the PRC earned US$4 billion per year from Hong Kong and $6 billion in foreign exchange. Rebekah C. Cockram, *Cession and Retreat: Negotiating Hong Kong's Future, 1979–1984*, thesis, University of North Carolina at Chapel Hill, 2018.

[30] Joseph S. Cheng, "The Future of Hong Kong: Surveys of the Hong Kong People's Attitudes," *China Journal*, Vol. 12, July 1984.

[31] Yahuda, 1993.

[32] Cockram, 2018, p. 10.

Granting Independence to Colonial Countries and Peoples applied.[33] Fearing that Hong Kong would gain status as an independent nation through these proceedings, the Chinese government stated in the letter that "[t]he settlement of the questions of Hong Kong and Macau is entirely within China's sovereign right and does not at all fall under the ordinary category of 'colonial territories.'"[34] Notably, this request was not taken up for debate in the UN General Assembly.[35] Instead, the request was brought up by the special committee for approval. When it was announced at the General Assembly, the motion was supported by Afghanistan, the USSR, and Mali, with only Venezuela, Fiji, and Sweden issuing concerns.[36] The motion was passed, and Hong Kong was removed from the list of territories named in the declaration.

Great Power Alignment

During World War II, the United States attempted to convince the UK to cede Hong Kong to the democratic Chinese government, but this pressure waned following the communist Chinese victory in the Chinese Civil War.[37] Beginning in the 1950s, the United States began to treat Hong Kong as a "political show window" and "an outpost of the free world" in Asia to entice Asian observers to democratic and capitalistic values.[38] Despite this interest, the United States did not offer any public resistance to the removal of Hong Kong from the UN list of colonies.

Takeover

Significant discussions for the future of Hong Kong began with a series of meetings from 1979 to 1984, which culminated in the Sino-British Joint Declaration of 1984, formally beginning the process of ceding Hong Kong to the PRC. Initially, the UK sought to extend the lease, believing that China held a firmer interest in the economic opportunities of Hong Kong than in the land, but China was firmly unwilling to consider any scenario that did not include full control over Hong Kong.[39]

Regional and Great Power Reactions

Britain considered the possibility of retaining the permanently ceded territories and giving China only the leased land, but analysts found that the ceded territories could not operate independently because of the large amount of resources, citizens, and infrastructure that crossed between the two

---

[33] UN General Assembly Resolution 1514 (XV), Declaration on the Granting of Independence to Colonial Countries and Peoples, adopted on December 14, 1960; Patricia A. Dagati, "Hong Kong's Lost Right to Self-Determination: A Denial of Due Process in the United Nations," *NYLS Journal of International and Comparative Law*, Vol. 13, No. 1, 1992.

[34] Huang, 1972.

[35] Dagati, 1992.

[36] Dagati, 1992.

[37] Yahuda, 1993; Maxwell, 1995.

[38] Fong, 2022, pp. 256–257.

[39] Cockram, 2018.

sections daily.[40] In the wake of waning international influence, the UK—which was eager to maintain the moral high ground and was concerned that the PRC would seek to reclaim Hong Kong at any cost—agreed to pass ownership of the entirety of Hong Kong to the PRC.[41]

However, British officials noted that negotiations with China were marked by China's unwillingness to compromise, voicing concerns that China's assurances were akin to "a scrap of paper."[42] For Hong Kong residents, these discussions were met with frustration and fear because their interests and preferences were not consulted by either the UK or China throughout the negotiations. As one contemporary academic wrote,

> [T]he people of Hong Kong have not been, and cannot be, masters of their own fate. They were excluded from the Sino-British negotiations and are still pawns in the discord and struggle for control between two sovereign powers. The community has been engulfed in a deep sense of helplessness stemming from their inability to change decisions made by Beijing and London.[43]

These frustrations and alarm at a lack of representation were echoed by journalists and international human rights groups, who accusing British Prime Minister Margaret Thatcher of preferring to gain international clout by leaving the territory peacefully at the expense of the rights and future of the people of Hong Kong.[44] Additionally, criticism by the UK and China of the overall situation, which often leaked to the Hong Kong media, led to a devaluing of the local currency and a drop in the stock market.[45] These uncertainties led to a wave of Hong Kong citizens emigrating to other—usually democratic—nations.[46] Surveys of Hong Kong residents at the time indicated that most respondents preferred maintaining the status quo and did not wish to emigrate or become Chinese citizens.[47]

Despite the impression among the people of Hong Kong that the UK had excluded them, the UK had requested to include Hong Kong representatives in the Sino-British discussions. However, Chinese officials maintained the stark position that Hong Kong residents did not need a seat at the table because the will of China was the will of Hong Kong citizens.[48] China's primary interest throughout the process was the return of Hong Kong, which the PRC believed to be an indivisible

---

[40] According to British analysis, 60 percent of the population lived on the leased land—which also had all major water reservoirs, the majority of fresh vegetables and livestock, and the port. The land permanently owned by Britain held 67 percent of the educational institutions and 88 percent of the economic activity (letter from Haddon-Clave to Richard Clift, July 26, 1982, British National Archives, Foreign Commonwealth Office 21/2214, cited in Cockram, 2018).

[41] Hook, 1997; Chan, 1996; Joint Declaration of the Government of the United Kingdom of Great Britain and Northern Ireland and the Government of the People's Republic of China on the Question of Hong Kong, signed at Beijing, China, December 19, 1984.

[42] Maxwell, 1995, p. 1388.

[43] Chan, 1996.

[44] Dagati, 1992.

[45] Maxwell, 1995.

[46] Top destinations for Hong Kongers included Canada, the United States, Australia, New Zealand, Singapore, and the UK (Chan, 1996).

[47] In a 1982 survey, 70 percent of respondents indicated wanting to maintain the status quo, 15 percent wanted to become British, and 4 percent wanted to go under PRC rule. In another survey, 69 percent wanted to maintain the status quo (Cheng, 1984).

[48] Yahuda, 1993.

part of Chinese territory that had been unfairly and illegally occupied by the British. The PRC placed significant pride in regaining Hong Kong, with the Chinese Communist Party (CCP) vice chair Deng Xiaoping making reunification a key part of his political strategy both domestically and internationally.[49] The PRC believed that the unequal treaties were signed by the previous government and made without the current leadership's consent, which rendered them invalid.[50] Official Chinese rhetoric surrounding the exchange focused on the invalid nature of the initial treaty, arguments that Britain had already ceded sovereignty, beliefs that the citizens of Hong Kong were Chinese, and threats that, if a formal agreement to cede the territory was not reached by a September 1984 deadline, China would made the decision alone.[51]

As a result of the UK's unwillingness to push the issue and China's firm stance, an agreement was signed in December 1984 to formally pass the entirety of Hong Kong from British to Chinese control: The Sino-British Joint Declaration established the territory of Hong Kong as a special administrative region (SAR) under a "one country, two systems" policy, which would ensure that the colony would maintain a "high degree of autonomy" for 50 years from the signing date.[52] The SAR would be governed according to the Basic Law, drafted by the PRC, which would protect the rights of Hong Kong residents and maintain freedom of the press, assembly, religion, travel, and association. Finally, the UK and China would establish a Joint Liaison Group to monitor the transition of governments and handle disputes.[53]

Aftermath

Following the Joint Declaration of 1984, the process of dissolving the colonial administration and placing Hong Kong under Chinese control began. These efforts were brought to an abrupt halt following student protests in China, which were violently suppressed in Tiananmen Square in 1989.[54] This event reversed Hong Konger interest in reunification and led to a major rift in the Sino-British cooperation to transfer the colony. British officials refused to work with Chinese officials because they suspected that China's efforts to protect human rights were insincere; at the same time, Chinese representatives suspected that the UK was purposefully sabotaging the handoff to erode China's

---

[49] Yahuda, 1993.

[50] Permanent Representative of China, letter to the UN addressed to the Chairman of the Special Committee on Decolonization, Annex 1, UN document A/AC.109/396, March 8, 1972.

[51] Cockram, 2018, p. 43; Maxwell, 1995; Robin M. White, "Hong Kong, Nationality and the British Empire: Historical Doubts and Confusions on the Status of the Inhabitants," *Hong Kong Law Journal*, Vol. 19, 1989.

[52] Joint Declaration of the Government of the United Kingdom of Great Britain and Northern Ireland and the Government of the People's Republic of China on the Question of Hong Kong, 1984, p. 61.

[53] Joint Declaration of the Government of the United Kingdom of Great Britain and Northern Ireland and the Government of the People's Republic of China on the Question of Hong Kong, 1984.

[54] The official number of deaths is listed as 100; however, some sources list it as high as 20,000, while many hover in the low thousands. The protests began regarding political representation, job prospects, corruption, and other complaints against the government and were led primarily by students (Wendy Dullea Bowie, "The Effect of the Tiananmen Square Massacre upon Negotiations for the Draft Basic Law of the Hong Kong Special Administrative Region Comment," *Penn State International Law Review*, Vol. 8, No. 2, 1990).

authority.[55] This resulted in parallel, competing efforts to solidify democratic and legal rights by the British, on the one hand, and to ensure central authority and economic stability by the Chinese, on the other hand.

Regional Alignment

For the people of Hong Kong, the events in Tiananmen Square that demonstrated the PRC's hostile and violent response to democratic initiatives indicated that the position of Hong Kong as an SAR—therefore, retaining a representative government apart from the rest of mainland China—would not be maintained by the PRC following the handover of the colony. A sentiment of bitterness and distrust toward both China and the UK arose among Hong Kongers along with a feeling that the UK had betrayed them.[56] One Hong Kong newspaper remarked that the working group that had aided in the transition had failed to accomplish anything beyond "alienat[ing] themselves from the Hong Kong population."[57] As a result, emigration out of Hong Kong rose, which led to a decline in wealth and a significant brain drain as wealthy and well-educated citizens moved to other nations.[58] Many sources indicated that Hong Kongers were frustrated because they saw the UK as willing to risk the lives and rights of its former colonial residents in order to gain status on the international stage.[59]

Despite Britain's unwillingness to formally protest the end of the lease, the events of Tiananmen Square changed its approach to the handover, and a scramble ensued to provide Hong Kong some level of protection and a precedent to defend the freedoms laid out in the agreements. In contrast to Britain's prior reluctance to establish democratic institutions, it began a series of social reforms and expanded democratic institutions.[60] Despite these efforts, critics noted that Britain gave local Hong Kong stakeholders just over a decade to organize themselves and develop democratic systems prior to the Chinese takeover, which was considered an insufficient amount of time to provide significant safeguards to Hong Kong residents.[61]

Although the UK was seeking to establish democratic institutions, the PRC was working to counter what it saw as a "recolonization" attempt and to solidify China's rule.[62] In reaction to both Tiananmen Square and the 1991 election, China began to fear a larger conspiracy and took steps to centralize control, including through the addition of clauses to the Basic Law concerning subversion.[63] As distrust between China and the UK grew, Chinese officials began questioning individual Hong

---

[55] Maxwell, 1995.

[56] Chan, 1996.

[57] Fanny Wong, "Battle to Believe in the PWC," *South China Morning Post*, October 19, 1994.

[58] In the early 1980s, approximately 20,000 people emigrated from Hong Kong per year. By 1987, this rose to 30,000 per year. In 1990, the year following the Tiananmen Square incident, emigration rose to 62,000 (Cait Murphy, "Hong Kong: A Culture of Emigration," *Atlantic Monthly*, Vol. 267, No. 4, April 1991; Chan, 1996).

[59] Yahuda, 1993; Wong, 1994; Ming K. Chan, "The Legacy of the British Administration of Hong Kong: A View from Hong Kong," *China Quarterly*, Vol. 151, September 1997.

[60] Granahan, 1992.

[61] Yahuda, 1993.

[62] Chan, 1996, p. 17.

[63] Yahuda, 1993; Maxwell, 1995.

Kong civil servants about their duties and plans, which some viewed as "an allegiance test."[64] Finally, China took steps to alter the direction of funds, prepare to hold new elections in 1997, and adjust education standards to conform with PRC education.[65]

Great Power Alignment

Internationally, many nations expressed concern about China's handling of Tiananmen Square and the precedent it set for its administration of Hong Kong, but little action was taken to address these concerns. The United States issued the United States–Hong Kong Policy Act in 1992, which allowed the United States to continue treating Hong Kong as separate from the PRC for economic issues and trade exports.[66] This act linked the future of Hong Kong's autonomy to the status of U.S.–Hong Kong relations, reserving the right for the U.S. president to suspend special treatment of Hong Kong if it were determined that Hong Kong was no longer able to operate independently.[67] Critics noted that this stance allowed China to continue benefiting from Hong Kong's economic structure while failing to indicate that the United States had an interest in the civil liberties of Hong Kong's citizens.[68] When the handover occurred, U.S. stakeholders expressed frustration at the lack of U.S. involvement in the negotiations and the handover itself, arguing that the United States' part as an "interested bystander" gave China free rein to renege on the commitments it had made with the UK.[69]

In 2020, the U.S. Department of State declared that Hong Kong was no longer autonomous because of increasing PRC control over Hong Kong; this declaration was formalized in "The President's Executive Order on Hong Kong Normalization," which suspended or eliminated aspects of the separate treatment of Hong Kong from China.[70] Additionally in 2020, the United States enacted the Hong Kong Autonomy Act, denouncing the PRC's treatment of Hong Kong as being in violation of the 1984 Joint Declaration and the Basic Law, and issued a series of sanctions on individuals and institutions that flouted these documents.[71] In 2021, the U.S. Department of the Treasury issued a formal business advisory, warning U.S. companies about the risks and considerations of doing business in Hong Kong, given the increase in PRC oversight.[72] Despite the United States' measures to denounce the PRC's growing control over Hong Kong, the United States still takes a primarily passive role in the issue. In 2024, U.S. stakeholders expressed frustration that

---

[64] Wong, 1994.

[65] These efforts were described by one journalist as follows: "[T]he assertiveness only serves to shatter the hopes that the promise of one country, two systems is a sincere and genuine one" (Wong, 1994).

[66] Fong, 2022.

[67] Public Law 102-383, United States–Hong Kong Policy Act of 1992, October 5, 1992.

[68] Fong, 2022; Mitch McConnell, "Hong Kong and the Future of China Three Years After the Tiananmen Square Massacre," Heritage Foundation, June 5, 1992.

[69] Jesse Helms, "We Will Take Up Hong Kong's Fight," *Wall Street Journal*, June 20, 1997.

[70] Jessica Donati, "U.S. Officially Declares That Hong Kong Is No Longer Autonomous," *Wall Street Journal*, May 27, 2020; Executive Order 13936, "The President's Executive Order on Hong Kong Normalization," Executive Office of the President, July 14, 2020.

[71] Public Law 116-149, Hong Kong Autonomy Act, July 14, 2020.

[72] Hong Kong Business Advisory, "Risks and Considerations for Businesses Operating in Hong Kong," U.S. Department of the Treasury, July 16, 2021.

the United States failed to take any action beyond a press release regarding the new Hong Kong national security legislation, which broadened the definition of security crimes in ways that drew criticism from human rights observers.[73] The new security law prompted the resignation of Canadian and British judges from Hong Kong courts; it also prompted Australia, Taiwan, and the UK to warn their citizens against travel to Hong Kong.[74]

Key Factors

The key factors that affected the course of events were as follows:

- *National pride:* The Chinese believed that the colony of Hong Kong was historically and rightfully an indivisible part of China. As a result, negotiations with the British were stilted, uncompromising, and plagued by suspicion on both sides that the other side would renege on its promises and did not understand the situation.
- *International decolonization:* The international pressure to divest remaining colonies and improve international standing made the UK reluctant to pursue an extended lease of Hong Kong. Additionally, the desire to behave cooperatively played a role in the lack of international pushback to China's request for control of the colony.
- *Ethical considerations:* Although most international observers initially supported China's administration of Hong Kong, human rights concerns following the Tiananmen Square massacre and subsequent reductions of the rights and protections of Hong Kong citizens have shifted the balance away from support of Chinese rule.

Conclusion

The handover of Hong Kong was done peacefully on July 1, 1997. Immediately following the handover, China marched the People's Liberation Army into Hong Kong and took control of the SAR. Despite concerns that British businesses would struggle under the PRC regime, many continued operating in Hong Kong following the transfer without significant disruption. After the PRC takeover, the percentage of citizens who were optimistic about the unification fell from 55 percent in 1997 to 23 percent by 2002.[75] Despite assurances by the PRC that all agreements with the UK would remain in place for at least 50 years, China has steadily and gradually eroded the Basic Law and the Universal Declaration of Human Rights through reinterpretations of the agreements, including notable reductions to Hong Kong freedoms in 2003, 2012, 2014, 2019, 2020, and 2024.[76] These

[73] Government of Hong Kong, "Safeguarding National Security Ordinance," March 19, 2024b; Human Rights Watch, "Hong Kong: New Security Law Full-Scale Assault on Rights," webpage, March 19, 2024.

[74] "Hong Kong's New Security Law Comes into Force amid Human Rights Concerns," Al Jazeera, March 23, 2024.

[75] Hong Kong Transition Project, "The First Five Years: Floundering Government, Foundering Democracy?" Hong Kong Baptist University, Government and International Studies, May 2002.

[76] Lindsay Maizland and Clara Fong, "Hong Kong's Freedoms: What China Promised and How It's Cracking Down," webpage, Council on Foreign Relations, last updated March 19, 2024; Kenji Kawase, "Hong Kong Passes Article 23 Security Law with Sweeping Powers," Nikkei Asia, March 19, 2024.

reductions have led to waves of Hong Konger protests, including the 2014 Umbrella Movement, sit-ins and occupations of government buildings, and street protests—some of which drew more than 1 million protesters.[77] As a result, the Chinese government has increased crackdowns over time. From the 2019 to 2024, the budget for the Hong Kong Police Force rose from 23.7 billion Hong Kong dollars (HK$) to HK$27.8 billion.[78] Additionally, the Hong Kong government added a new "nonrecurrent" line item of HK$8 billion to its 2020 budget and HK$5 billion to its 2023 budget for a "special fund to meet the expenditure for safeguarding national security."[79]

One long-term effect of Hong Konger criticism regarding Britain's handling of the 1997 handover was that, following Chinese government crackdowns in Hong Kong in 2019 and 2020, the UK implemented a new British Nationals (Overseas) visa for Hong Kong citizens looking to emigrate to Britain, creating an easier path to British emigration.[80] Although China has continued reducing freedoms in Hong Kong, it maintains that Hong Kong is a shining example of an SAR and a model for unification of previously held Chinese territories under the "one country, two systems" model.[81] Such efforts have failed to entice Taiwan into following Hong Kong's example.

Although initially friendly, negotiations on the future of the colony of Hong Kong soured following China's violent suppression of protests. Although the transfer of the entire colony occurred peacefully, Hong Kong citizens remain frustrated by their lack of representation throughout the process. China's position throughout negotiations was firm on the status of Hong Kong as rightful Chinese territory and the PRC as the formal mouthpiece of Chinese (including Hong Konger) citizens, while the UK was most concerned about its international image. International stakeholders initially supported the removal of colonial structures in Hong Kong, but subsequent losses of freedom have led the UK and the United States to issue statements and enact policies condemning Chinese administrative methods.

The cession of Hong Kong offers an example of the factors that might play a role in the peaceful incorporation of territory that many in the international community recognized as rightfully a PRC territory. These factors include both short- and long-term perceptions of the nation responsible for the territorial revision. Because the PRC has generally not been seen as abiding by its commitments prior to the cession and as violating the rights and protections of Hong Kong citizens, the handover of Hong Kong is seen by many as a failed model of peaceful cession.[82] In other cases of peaceful incorporation, similar factors would likely play a role in determining international responses. Namely,

---

[77] Austin Ramzy and Mike Ives, "Hong Kong Protests, One Year Later," *New York Times*, last updated June 22, 2021.

[78] Government of Hong Kong, "Summary of Expenditure Estimates," 2021; Government of Hong Kong, "Summary of Expenditure Estimates," 2024a.

[79] This special fund is separate from the annual Hong Kong Police Force budget (Government of Hong Kong, 2021; Government of Hong Kong, "Summary of Expenditure Estimates," 2023; Maya Wang, "Hong Kong Repression's True Cost: Ominous 'National Security' Budget Increase Spotlights Growing Abuse," Human Rights Watch, February 24, 2021).

[80] Heather Rolfe and Thomas Benson, *From HK to UK: Hong Kongers and Their New Lives in Britain*, British Future, Department for Levelling Up, Housing, and Communities, November 2023.

[81] People's Republic of China, *The Taiwan Question and China's Reunification in the New Era*, Taiwan Affairs Office of the State Council and State Council Information Office, August 2022.

[82] As discussed previously, there were disagreements over the nature of what was promised during the cession. Because our report focuses on third-party responses, the fact that much of the international community sided with the Hong Kongers is critical to our discussion.

whether the incorporating nation abided by its commitments prior to the transition and adhered to tenets of human rights would play a key role in determining whether other states saw the incorporating nation's actions as representing a threat to the region in need of balancing action. These concerns would, as was the case in Hong Kong, be weighed against economic ties and the potential for both gains and losses a result of regional and international state responses.

Chapter 3

The Soviet Invasion of Czechoslovakia, 1968

On August 20, 1968, the Soviet Union led Warsaw Pact troops in an invasion of Czechoslovakia, swiftly gaining control of the country.[83] In the decades leading up to the invasion, political tensions between Moscow and Prague had escalated as Czechoslovak politicians and military leaders initiated reforms to slowly distance their country from Moscow leadership.[84] Dissent from a conservative, pro-Soviet government was considered a threat to Moscow because of the heavy reliance on the Czechoslovak army among the Warsaw Pact parties to mobilize against any NATO aggression.[85] Czechoslovak defiance of Moscow's leadership also provoked fears that rebellion in other countries would follow and organize a larger insurrection in the Eastern Bloc.[86] Ultimately, Moscow invaded and installed a new government in Prague, successfully stifling anti-Soviet reforms.[87]

Fears of an emboldened Soviet Union reverberated across the communist world, with some states shifting away from Moscow because they feared an invasion that would force Moscow's vision of communism on smaller states. However, it was largely Eastern Bloc countries that experienced this threat of successive invasion. Czechoslovakia was far behind the Iron Curtain, and the Soviet invasion was swift, which limited the opportunity for Western intervention. In any event, noncommunist countries feared that direct action would upset the Cold War status quo and escalate tensions, superseding concerns about an emboldened Soviet Union. More importantly, Soviet head of state Leonid Brezhnev clearly signaled that his eponymous doctrine, which imposed a unified vision emanating from Moscow, was limited to current communist countries.[88] This simultaneously reduced fears of Russian aggression toward Western states and exacerbated the ongoing Sino-Soviet split.

The 1968 Soviet invasion of Czechoslovakia is a pertinent case study for analyzing future territorial revisions because it underscores how geopolitical considerations can create conditions for

---

[83] Office of the Historian, Foreign Service Institute, "Soviet Invasion of Czechoslovakia, 1968," webpage, U.S. Department of State, undated-a.

[84] "From Spring, into a Long Winter's Night: The Czechoslovakian Crisis of 1968," Cryptologic Almanac 50th Anniversary Series, DOCID: 3519758, last reviewed February 28, 2003; U.S. Central Intelligence Agency, "The CIA and Strategic Warning: The 1968 Soviet-Led Invasion of Czechoslovakia: An Overview," September 1, 2009.

[85] "From Spring, into a Long Winter's Night: The Czechoslovakian Crisis of 1968," 2003.

[86] Office of the Historian, Foreign Service Institute, undated-a.

[87] Office of the Historian, Foreign Service Institute, undated-a.

[88] "Unofficial Enunciation of the 'Brezhnev Doctrine,'" in Jaromír Navrátil, Antonín Benčík, Václav Kural, Marie Michálková, and Jitka Vondrová, eds., *The Prague Spring 1968: A National Security Archive Documents Reader*, Central European University Press, [September 26, 1968] 1998.

the invasion of a state without substantial international resistance. Following World War II, the delicate equilibrium between the Soviet Union and Western powers allowed the Soviet invasion of Czechoslovakia to proceed largely uncontested. A similarly precarious balance persists between many nations today and will likely be a significant consideration when nations weigh the benefits and drawbacks of becoming involved in potential conflicts. Future conflicts could again stifle willingness to engage in supporting smaller states for fear of escalation, which might lead nations to favor de-escalation—as was the case during the Warsaw Pact invasion of Czechoslovakia.

Historical Context

Regional Alignment

Following World War II, the Soviet Union installed a communist government in Czechoslovakia, a once robust democracy, and thwarted Czechoslovak attempts to join the U.S. Marshall Plan—the U.S. economic assistance program to help Europe rebuild after World War II.[89] The installation of a communist government was part of the Soviets' strategy to consolidate influence in the region through the establishment of the Warsaw Pact and counterbalance the burgeoning partnerships within NATO.[90] The Warsaw Pact was particularly reliant on Czechoslovakia's role in Soviet military plans, specifically to command the front alongside Poland in the case of war.[91] The reduction of Soviet ground forces in the 1960s also reinforced the reliance on the Czechoslovak army in the Warsaw Pact.[92]

By the mid-1960s, the Warsaw Pact was seen by smaller communist states as a conduit for more autonomy.[93] De-Stalinization in Russia had led to instability in Hungary and Poland. Yet, Czechoslovakia remained stable under the conservative communist government.[94] However, as one of the most industrially developed countries in Eastern Europe, Czechoslovakia had suffered substantially under communist rule, and the public began to blame the antiquated party hierarchy.[95] The appointment of Alexander Dubček as First Secretary marked a turning point because he embarked on a bold agenda, known as the Prague Spring, to liberalize Czechoslovak politics and the economy.[96]

Dubček's reforms allowed free speech, noncommunist parties, and democratic elections.[97] Dubček reinstated public political debates and freedom of travel, ended censorship, and removed conservative

89 Office of the Historian, Foreign Service Institute, undated-a.

90 U.S. Central Intelligence Agency, 2009.

91 U.S. Central Intelligence Agency, 2009.

92 U.S. Central Intelligence Agency, 2009.

93 U.S. Central Intelligence Agency, 2009.

94 Office of the Historian, Foreign Service Institute, undated-a.

95 U.S. Central Intelligence Agency, 2009.

96 U.S. Central Intelligence Agency, 2009.

97 U.S. Central Intelligence Agency, 2009.

leadership in the army, police, and Ministry of Defense.[98] Dubček even proposed opening diplomatic and trade relations with Western countries.[99]

Meanwhile, Brezhnev pushed his own reforms that shifted his totalitarian regime to one of more tolerance.[100] Brezhnev's reforms inspired hope—from Dubček and the Czechoslovak public—that Dubček might be able to secure autonomy for Czechoslovakia.[101] Indeed, Soviet leadership was concerned that tensions within Czechoslovakia in regard to its shrinking economy under communism could affect the reliability of Prague's military contributions to the Warsaw Pact.[102] However, Soviet leaders became fearful that if Czechoslovakia carried reforms too far, other Eastern countries and Soviet republics might follow suit and lead to a rebellion against Moscow's leadership.[103] In the nine months leading up to the invasion, Soviet leaders graduated from voicing concerns about the reforms to threatening the Czechoslovak government.[104] A series of dialogues between Moscow and Prague were an attempt to stave off direct Soviet action. However, in the end, the compromises that were obtained were ultimately deemed insufficient, and Moscow intervened militarily and established a more-conservative, pro-Soviet government.

Great Power Alignment

The West had no intention of intervening in the Prague Spring because it was viewed as an issue internal to the Warsaw Pact.[105] In July 1968, strict orders were given to U.S. armed forces in West Germany to halt all activity on the Czechoslovak border that could be interpreted by the Soviets as support for the Prague Spring.[106] The United States also reduced its military presence in Europe to further distance itself from the developing conflict and to avoid appearing to support Czechoslovakia.[107]

However, the Prague Spring exacerbated the Sino-Soviet split, escalating tensions between China and the USSR. To bolster his domestic legitimacy and gain support among Eastern Bloc countries, CCP chair Mao Zedong accused the Soviets of betraying the global communist movement. Chinese propaganda specifically accused Moscow of "big power chauvinism" over its handling of Czechoslovakia.[108] Responses to Czechoslovakia's movement varied in Eastern Europe as well. The

[98] "From Spring, into a Long Winter's Night: The Czechoslovakian Crisis of 1968," 2003.

[99] Anna J. Stoneman, "Socialism with a Human Face: The Leadership and Legacy of the Prague Spring," *History Teacher*, Vol. 49, No. 1, November 2015.

[100] Stoneman, 2015.

[101] U.S. Central Intelligence Agency, 2009.

[102] U.S. Central Intelligence Agency, 2009.

[103] Office of the Historian, Foreign Service Institute, undated-a.

[104] U.S. Central Intelligence Agency, 2009.

[105] U.S. Central Intelligence Agency, 2009.

[106] Jiri Valenta, "The Bureaucratic Politics Paradigm and the Soviet Invasion of Czechoslovakia," *Political Science Quarterly*, Vol. 94, No. 1, Spring 1979.

[107] Office of the Historian, Foreign Service Institute, undated-a.

[108] Filip Šebok, "Ideology and Geopolitics: Chinese Reaction to the 1968 Invasion of Czechoslovakia," Central European Institute of Asian Studies, August 21, 2018.

Soviets' heavy-handed response earned Czechoslovakia some sympathy from such countries as Hungary, which had been subject to similar treatment a decade earlier.[109] The communist dictators of Poland and East Germany, however, watched the Prague Spring with particular fear.[110] In June, access to East Germany was used as an additional factor for the Soviet need to sustain Warsaw Pact troops on Czechoslovak soil.[111] East Germany was also concerned with maintaining domestic stability in the case that Czechoslovakia continued to liberalize.[112] Overall, it was the need to potentially cope with unstable conditions in Poland and East Germany—rather than a threat from the West—that pushed Moscow to suppress the Prague Spring.[113]

Events

A week prior to the invasion, Brezhnev's troops were readied for action in a final attempt to pressure Dubček to take severer measures in response to Czechoslovak media disparaging the Soviet Union.[114] However, Dubček insisted that no immediate action nor any timeline was stipulated in the agreements made in July.[115] Dubček argued that anti-Soviet sentiment would be resolved at a Central Committee plenum that was yet to be scheduled.[116] Transcripts between Brezhnev and Dubček reveal that Dubček refused to provide a plan to resolve the Soviet concerns. Instead, Dubček ridiculed Brezhnev and told him to "take the measures you regard as appropriate. That's your affair."[117] Brezhnev agreed that he undoubtedly would.[118]

On August 20, 1968, the Soviet Union surrounded Czechoslovakia with troops in Hungary, Poland, East Germany, and Bulgaria for a "Warsaw Pact military exercise."[119] Russian authorities arrested Dubček and four other leaders of the Communist Party of Czechoslovakia in the night, took them to Moscow, and forced them to "approve" the pending military occupation.[120] Moscow sent 200,000 Warsaw Pact troops and 2,000 tanks into Czechoslovakia to brutally suppress the Prague

---

[109] U.S. Central Intelligence Agency, "'Czechoslovakia: The Dubcek Pause,' U.S. Intelligence Assessment of the Crisis," in Jaromír Navrátil, Antonín Benčík, Václav Kural, Marie Michálková, and Jitka Vondrová, eds., *The Prague Spring, 1968: A National Security Archive Documents Reader*, Central European University Press, [June 1968] 2006.

[110] "From Spring, into a Long Winter's Night: The Czechoslovakian Crisis of 1968," 2003.

[111] U.S. Central Intelligence Agency, 2006.

[112] Valenta, 1979.

[113] Valenta, 1979.

[114] Leonid Brezhnev, "Transcript of Leonid Brezhnev's Telephone Conversation with Alexander Dubcek, August 13, 1968," in Jaromír Navrátil, Antonín Benčík, Václav Kural, Marie Michálková, and Jitka Vondrová, eds., *The Prague Spring, 1968: A National Security Archive Documents Reader*, Central European University Press, [August 13, 1968] 2006, p. 172.

[115] Brezhnev, 2006.

[116] Brezhnev, 2006.

[117] Brezhnev, 2006.

[118] Brezhnev, 2006.

[119] Office of the Historian, Foreign Service Institute, undated-a.

[120] Antonín Benčík et al., "Report by a Czechoslovak State Security (StB) Official on the Arrest of Dubček and Other Members of the CPCz CC Presidium," in Jaromír Navrátil, Antonín Benčík, Václav Kural, Marie Michálková, and Jitka Vondrová, eds., *The Prague Spring, 1968: A National Security Archive Documents Reader*, Central European University Press, [August 21, 1968] 2006; Stoneman, 2015.

Spring movement, killing 186 civilians and seriously wounding 362.[121] In just one day, the Soviet Union decisively took control of Czechoslovakia's major cities, means of communication, and transportation lines.[122]

Dubček nominally remained in leadership until April 1969, when he was replaced by First Secretary Gustáv Husák.[123] Fearful of receiving the same brutal response as the Hungarian uprisings received in 1956, Dubček had sustained his position for eight months by demonstrating loyalty to Moscow and Czechoslovakia's intention to remain in the Eastern Bloc.[124]

Regional Reactions

Many surrounding Warsaw Pact states saw Czechoslovakia's liberalizing reforms as threats to their own communist governments.[125] The Soviet Union, East Germany, Poland, Hungary, and Bulgaria all contributed troops for the invasion.[126] Though the invasion was successful, a small resistance persisted for one year while the Soviet Union struggled to install a stable government.[127]

Although many Warsaw Pact countries signaled support of the invasion, other communist countries, including China, Yugoslavia, and Romania, condemned it.[128] Although Romania remained steadfast in sustaining its position in the Eastern Bloc, the Romanian President Nicolae Ceaușescu declared that the invasion was a "grave error and constituted a serious danger to peace in Europe for the prospects of world socialism."[129] However, in private conversations with the U.S. ambassador to the UN, the Romanian delegation asserted that "the Czechoslovaks had gotten themselves into an impossible position, both domestically and internationally, and had not known what to do."[130] The U.S. ambassador then assured the delegation that a Russian invasion of its country would elicit a much stronger response from the West.[131]

---

[121] Stoneman, 2015.

[122] Office of the Historian, Foreign Service Institute, undated-a.

[123] U.S. Central Intelligence Agency, 2009.

[124] U.S. Central Intelligence Agency, 2009.

[125] "From Spring, into a Long Winter's Night: The Czechoslovakian Crisis of 1968," 2003.

[126] Office of the Historian, Foreign Service Institute, undated-a; "Report to the British Cabinet Summing Up the Situation in Czechoslovakia," The National Archives (United Kingdom), CAB 128/43, August 1968. Some sources claim that East Germany participated on only a planning level (Fred Frommer, "When Soviet-Led Forces Crushed the 1968 'Prague Spring,'" History, March 14, 2022).

[127] Office of the Historian, Foreign Service Institute, undated-a.

[128] "Chinese Reactions to the Invasion of Czechoslovakia," *Studies in Comparative Communism*, Vol. 2, No. 1, 1969; Dennis Deletant, "New Evidence on Romania and the Warsaw Pact, 1955–1989," Wilson Center, July 7, 2011; U.S. Central Intelligence Agency, "The Invasion of Czechoslovakia: Impact on Yugoslavia," Office of National Estimates, October 9, 1968.

[129] Sophie May, "Ceaușescu and the Soviet Invasion of Czechoslovakia: The Early Years of Dealing with a Dictator," webpage, Association for Diplomatic Studies and Training, undated. See also Harry Barnes, interview with Charles Stuart Kennedy, Association for Diplomatic Studies and Training Foreign Affairs Oral History Project, April 25, 2001.

[130] "Memorandum of Conversation: Secretary's Delegation to the Twenty-Third Session of the United Nations General Assembly," *Foreign Relations of the United States, 1964–1968*: Vol. 17, *Eastern Europe*, October 1, 1968.

[131] "Memorandum of Conversation: Secretary's Delegation to the Twenty-Third Session of the United Nations General Assembly," 1968.

Given the swift nature of the takeover, Czechoslovakia and much of the Western world were caught by surprise and watched Soviet actions with nervousness.[132] NATO allies preferred lessening tensions with the Soviet Union to intervention.[133] The USSR blocked efforts within the UN to condemn the invasion.[134] Ultimately, Western Europe settled on characterizing the invasion as a part of changes necessary to create stability within the Warsaw Pact countries.[135]

Great Power Reactions

Given Washington's escalating involvement in Vietnam, the United States refrained from intervening in yet another international situation.[136] Instead, the United States quickly condemned the Soviet Union for its interventionist actions and called for the Soviets to contain their actions to Czechoslovakia.[137]

China also quickly condemned the invasion, comparing it with Nazi Germany's invasion of Czechoslovakia in 1939 and arguing that the Soviet justification was similar to Nazi Germany's claim to be "defending peace."[138] China also condemned Dubček for not deploying the Czechoslovak people to fight the occupation. Chinese statements and propaganda were conflicting, with some claiming that the invasion was carried out to prevent Czechoslovakia from turning capitalist like the United States.[139] Despite the conflicting narrative from China, it was clear that Beijing blamed the United States for the invasion.[140] China claimed that the invasion was the result of collusion between the "Soviet revisionist renegade clique" and U.S. imperialism.[141]

Aftermath

Dubček's successor Husák again aligned Czechoslovakia with the Soviet Union, reestablishing censorship and preventing freedom of movement.[142] Ultimately, the invasion was successful in delaying the splintering of the Eastern Bloc.[143] Escalating tensions with the West eased as the Soviet Union imposed stability in Eastern Europe. However, other communist countries feared an emboldened

---

[132] Office of the Historian, Foreign Service Institute, undated-a.

[133] Office of the Historian, Foreign Service Institute, undated-a.

[134] Office of the Historian, Foreign Service Institute, undated-a.

[135] U.S. Central Intelligence Agency, 2009.

[136] Office of the Historian, Foreign Service Institute, undated-a.

[137] Office of the Historian, Foreign Service Institute, undated-a.

[138] "Chinese Reactions to the Invasion of Czechoslovakia," 1969, p. 115.

[139] Šebok, 2018.

[140] "Chinese Reactions to the Invasion of Czechoslovakia," 1969.

[141] "Chinese Reactions to the Invasion of Czechoslovakia," 1969, p. 115; Šebok, 2018.

[142] Office of the Historian, Foreign Service Institute, undated-a.

[143] Office of the Historian, Foreign Service Institute, undated-a.

USSR. In particular, China worried that the same Brezhnev Doctrine justifying the 1968 invasion would be used against it.[144]

Regional Alignment

The Soviet Union claimed that it aimed to strengthen socialism so that, to preserve world order and balance, not too much power lay in the hands of the Western world. The Soviet Union believed that all socialist countries had an obligation to protect their mutual security because damage to socialist positions could disturb the established international balance with Western countries.[145] At the 23rd UN General Assembly in the fall of 1968, the Soviet delegation argued that it had no choice but to resort to armed conflict after diplomacy failed.[146]

Instead, Soviet intentions to intervene only in other socialist state affairs reduced threat perceptions in Western Europe, which the Soviets considered to be outside their area of concern. According to the Soviet Union, however, a weakened socialism throughout the world would increase the danger of the outbreak of a new world war.[147]

In the name of preventing another world war, the Soviet Union was prepared to sign a document prohibiting the use of atomic weapons.[148] At the 23rd UN General Assembly, the Soviet Union also called for an end to the arms race and the beginning of disarmament. However, rather than "dry[ing] up the channels of proliferation of nuclear weapons," Moscow believed that a more stable world order would arise instead by granting access to these weapons and dispersing them to all or almost all states.[149] The Soviet Union pointed to the concentration of military presence and weapons in Europe, with the recent history of Nazi Germany as proof of the need to disseminate the geographic concentration of arms.[150]

Great Power Alignment

Although Western powers did not intervene in the invasion, the UN Security Council (UNSC) made several efforts to condemn the attacks; however, the Soviet Union vetoed these attempts.[151] The UN General Assembly blamed the USSR and its four Warsaw Pact allies for the events that had "poisoned" geopolitics.[152] However, Soviet tensions with the United States began to ease in the invasion's aftermath. Washington interpreted the Brezhnev Doctrine as applying only to communist

---

144 Office of the Historian, Foreign Service Institute, undated-a.

145 UN General Assembly, "Twenty-Third Session Official Records," 1,679th Plenary Meeting, at 10:30 a.m., New York, October 3, 1968.

146 UN General Assembly, 1968.

147 UN General Assembly, 1968.

148 UN General Assembly, 1968.

149 UN General Assembly, 1968.

150 UN General Assembly, 1968.

151 Office of the Historian, Foreign Service Institute, undated-a.

152 UN General Assembly, 1968.

countries. In this view, Moscow was attempting merely to consolidate its power in countries it already dominated rather than expanding Soviet influence into Western Europe.[153]

However, Beijing feared that the Soviets would use the doctrine to invade China.[154] As a result, the presence of a Soviet aircraft in Chinese airspace in September 15, 1968, was interpreted as an intentional "intrusion."[155] These tensions further intensified the hostility between China and the Soviet Union to the verge of "full-scale war" as the Soviet Union replaced the United States as China's primary enemy.[156] This had the secondary effect of reducing tensions between China and the United States, which both became concerned about the Soviet Union's intentions toward China.[157] This laid the foundation for U.S. diplomacy under Richard Nixon and Henry Kissinger that split the Soviets and Chinese.

Key Factors

The key factors that affected the course of events were as follows:

- *Cold War tensions:* The international community feared that intervention in the Soviet invasion would upset the status quo and escalate tensions. This fear seemed to prevail over concerns about an emboldened Soviet Union.
- *Waning influence:* Russia feared that Czechoslovakia was shifting away from its sphere of influence as a result of a series of reforms that favored more-democratic governance. Russia's invasion successfully restored Soviet influence over Czechoslovak politics and instilled fear in other Soviet bloc countries, recementing Soviet power in the region.
- *Changing threat perceptions:* The Soviet intervention in a fellow communist country changed the threat perception among both communist and noncommunist states. By making it clear through the Brezhnev Doctrine that the Soviets were concerned primarily about imposing a unified vision of communism in existing communist states rather than aggressive actions toward the democratic states, the West became less concerned about Soviet aggression while communist states began to see a greater threat. This had the effect of splintering the communist world.

Conclusion

Although Czechoslovakia may have once represented hope for a more liberal Eastern Bloc, its neighboring countries ultimately sided with Moscow to ease domestic unrest and prevent potential retaliation from the Soviet Union. The invasion succeeded in sustaining Soviet preeminence in Eastern Europe, but it only did so temporarily. The Soviet Union cited the same doctrine to justify

[153] Office of the Historian, Foreign Service Institute, undated-a.

[154] Office of the Historian, Foreign Service Institute, undated-a.

[155] Šebok, 2018.

[156] Šebok, 2018.

[157] Šebok, 2018.

invasion anywhere communist governments were threatened, and China feared that it could be next. Following the Czechoslovakia incursion, the Soviet Union did not hesitate to threaten the use of force against other socialist countries. In doing so, however, it galvanized a counterbalancing coalition, led by China, and further splintered the international communist movement. The United States condemned the Soviets' aggressive behavior but did not directly intervene because it feared escalation.[158] Instead, Washington seized on the cracks created within the Eastern Bloc by the USSR's actions, further deepening the wedge between Moscow and Beijing in the Sino-Soviet split.

This case study illustrates how one invasion can embolden a powerful state intent on forcing its vision on smaller neighbors. If an aggressive state succeeds in annexing additional territory, it may become emboldened to pursue further territorial ambitions in other contentious areas. This could particularly be the case if outside powers, such as the United States, do not intervene, which could be seen as giving carte blanche to the aggressor to strengthen its sphere of influence. Thus, smaller states may perceive a need to side with the growing power to avoid the same fate because they cannot trust that an outside power will be there to assist in the protection of their sovereignty.

[158] Office of the Historian, Foreign Service Institute, undated-a.

Chapter 4

Japanese Invasion of Manchuria, 1931

Prior to 1931, Japan had long enjoyed a quasi-colonial presence in northeast China.[159] However, as threats to Japan's interests in the region grew during the decade following World War I, some Japanese policymakers argued that their country needed a more permanent footprint in the area. On September 18, 1931, fighting between Japanese and Chinese troops broke out in response to reports of an explosion on a Japan-owned railway line in northeast China. This incident was a false flag staged by Japanese forces, who proceeded without imperial authorization to conquer and occupy the rest of northeast China for the following three months.

Japan hoped to keep the conflict bilateral, negotiating only and directly with the Chinese government. Tokyo feared that the other great powers would force it to relinquish some of its conquests in line with League of Nations' norms and the Kellogg-Briand Pact.[160] For their part, Washington, London, and Paris sought a mediated solution to contain the conflict. However, the weakness and fragmentation of international institutions—particularly the league—slowed great power coordination and their ability to force Japan to accept any decisions. The league eventually determined that Tokyo had acted aggressively and that any future administration of the region should be aligned with Chinese sovereignty.

The great powers consistently refused to accord Japan international status in line with Tokyo's expectations, which led to the collapse of the treaty port system and Japan's subordinate status in the Washington Naval Treaty.[161] The league's decision on Manchuria was the final straw that led Japan to withdraw from the organization. The inability of the great powers to maintain the international order allowed Tokyo to unilaterally pursue its territorial goals. The seizure of Manchuria posed significant risk of spillover, but primarily to China. The United States, Britain, and France were distracted by the Great Depression, and revanchist Germany further limited any appetite for intervention. Japan eventually seized the opportunity for further conquest, which resulted in the Second Sino-Japanese War (1937–1945).

This case provides context for situations in which the international community initially takes a strong stance in a potential territorial dispute but fails to maintain its hard line over time. In such a situation, although the instigating state is not provided with the opportunity to further its ambitions, the authority of the international institution that failed to enforce its decision is also weakened. By not

[159] We use the more neutral term *northeast China* rather than *Manchuria* to describe the region where Japan's possessions were located. The term *Manchuria* is an exonym that some scholars avoid because of its associations with Japanese colonialism. For treatment of this debate, see Mark C. Elliott, "The Limits of Tartary: Manchuria in Imperial and National Geographies," *Journal of Asian Studies*, Vol. 59, No. 3, August 2000.

[160] Kellogg-Briand Pact, signed at Paris, August 27, 1928.

[161] Rohan Mukherjee, *Ascending Order: Rising Powers and the Politics of Status in International Institutions*, Cambridge University Press, 2022; Steven Ward, *Status and the Challenge of Rising Powers*, Cambridge University Press, 2017.

recognizing Japanese authority to make territorial-revision decisions independently, the league initially contained the conflict but, later, became distracted by other affairs. As a result, Japan began a more aggressive campaign for territorial acquisition. Such shifting priorities eroded the authority of the league and allowed Japanese territorial revision.

Historical Context

Regional and Great Power Alignment

By the start of the 20th century, each of the world's great powers had secured a quasi-colonial presence in China under an arrangement known as the *treaty port system*: Although the specifics of each power's presence varied, the great powers typically enjoyed privileges ranging from trade rights to territorial concessions. These privileges were the result of treaties that took advantage of China's weakened state.[162]

Japan's possessions in China grew in 1905 when, following its victory in the Russo-Japanese war, Japan gained Russia's possessions in northeast China (also known as Manchuria). These possessions included a lease on the Liaodong Peninsula, a Russian-built railway, and the land and key towns adjacent to the railway. Japan would use these possessions as a foothold to grow its influence in northeast China over the next 26 years.[163]

Although the great powers jockeyed for influence in China, they had a common interest in seeing the treaty port system continue. For example, in 1900, a coalition of eight great powers invaded northern China to safeguard their citizens and possessions against the Boxer Rebellion.[164] However, the end of World War I left the treaty port system's beneficiaries less unified than before. Both sides saw their influence wane as a result of the war, regardless of victory or defeat in Europe. All sides emerged from the war hugely in debt, financially and demographically ruined, and lacking in any moral authority. Two stakeholders, France and Germany, had seen their influence in China wane because of the war. Another former stakeholder, the Russian Empire, had transformed from an expansionist force into an avowedly anti-imperialist Soviet Union after 1917. The remaining stakeholders, the United States and the UK, looked unfavorably on Japan's desire for greater privileges within the treaty port system (although they tolerated Japan's acquisition of Germany's possessions in China). Moreover, the U.S. and Australian decision to exclude a racial equality clause from the League of Nations charter convinced some Japanese policymakers that the Western powers would never grant an equal stake in the international system to an Asian nation. This suspicion would return to the forefront of Japanese foreign policy debates following the invasion of northeast China.[165]

Nevertheless, in the years immediately following World War I, the great powers were able to draft several treaties to manage disagreements that might have otherwise marked the end of the treaty port

---

162 W. G. Beasley, *Japanese Imperialism: 1894–1945*, Oxford University Press, 1987, pp. 14–20.

163 Louise Young, *Japan's Total Empire: Manchuria and the Culture of Wartime Imperialism*, University of California Press, 1998, p. 25.

164 Beasley, 1987, pp. 14–20.

165 Ward, 2017, pp. 100–129; Beasley, 1987, p. 157.

system. Under the Four-Power Treaty (1921),[166] the United States, Britain, France and Japan agreed to respect each other's territorial claims in Asia and pledged to resolve any territorial disputes diplomatically. Under the Washington Naval Treaty (1922),[167] Britain, the United States, and Japan agreed on a naval arms ratio. Finally, under the Nine-Power Treaty (1922),[168] the great powers reaffirmed their commitment to an Open Door policy toward China. These treaties sustained the treaty port system for a time.[169]

However, threats to the treaty port system continued to grow throughout the 1920s as the anticolonial movement in China led to protests, boycotts, strikes, and riots targeted at foreigners and foreign firms.[170] Amid this anti-colonial groundswell, the Guomindang and the recently founded CCP, the two main parties that vied for supremacy leading up to World War II and afterward, both committed themselves to reversing the treaties that underpinned the treaty port system.[171] Although violent opposition to the treaty port system had occurred before, these new protests posed a larger threat to the system because of the fractured contingent of great power stakeholders. The threat posed by the anti-colonial movement was worsened by the Chinese government's growing instability. For the treaty port system to work, there had to be a Chinese government strong enough to enforce the treaty rights of imperial powers, but the strength of the Chinese government was wavering in the face of challenges by warlords.

Japan's faith in the treaty port system began to erode, but it was not ready to give up on the system entirely; Japan may have taken these threats to the treaty port system seriously from the start, but it did not want to disrupt the parts of the system that it benefited from.[172] Therefore, Tokyo took a middle path between business as usual and outright military action: It began supporting Zhang Zuolin, a warlord who had come to control much of northeast China, by providing materiel and military advisers in exchange for various economic concessions.[173]

In 1926, Guomindang forces advanced north and began to threaten Zhang Zuolin, attempting to coerce him into reversing some of the concessions he had made to Japan.[174] While Japanese policymakers' fears grew as the Guomindang began to gain strength, Britain and the United States recognized the Guomindang as the legitimate government of China and entertained its proposals to revise the treaty port system.[175] In 1928, the Guomindang reached Beijing and forced Zhang Zuolin to retreat. Although policymakers in Tokyo were unwilling to separate northeast China from China

---

[166] Treaty Between the United States of America, the British Empire, France, and Japan, signed at Washington, D.C., December 13, 1921.

[167] Treaty Between the United States of America, the British Empire, France, Italy, and Japan, signed at Washington, D.C., February 6, 1922.

[168] Treaty Between the United States of America, Belgium, the British Empire, China, France, Italy, Japan, the Netherlands, and Portugal, signed at Washington, D.C., February 6, 1922.

[169] Ward, 2017, pp. 100–129.

[170] Beasley, 1987, p. 169.

[171] Beasley, 1987, p. 169.

[172] Beasley, 1987, p. 188.

[173] Beasley, 1987, p. 162.

[174] Beasley, 1987, p. 162.

[175] Beasley, 1987, pp. 171–172.

proper, elements of the Kwantung Army, Japan's garrison in the region, saw things differently. On June 4, 1929, a Kwantung Army colonel ordered the assassination of Zhang Zuolin in the hopes of provoking a Japanese occupation of northeast China. The plan backfired. Zhang Zuolin was succeeded by his son, Zhang Xueliang, who declared allegiance to the Guomindang in December 1929.[176]

Following the 1929 U.S. stock market crash, some Japanese policymakers came to believe that the treaty port system no longer served the empire's interests. Moreover, some were convinced that the empire's possessions in northeast China had become a "lifeline" for the Japanese economy, and some argued that this lifeline was worth protecting aggressively.[177] In 1931, the Kwantung Army drafted plans to occupy northeast China outright without Tokyo's approval.[178]

Events

After reports of an explosion on a Japanese-owned railway line on September 18, 1931, the Kwantung Army attacked Guomindang troops in northeast China. By the next morning, the Kwantung Army had captured the nearby city of Mukden (present-day Shenyang).[179]

Over the next three months, the Kwantung Army advanced and occupied the rest of northeast China—despite objections from the prime minister and foreign minister in Tokyo.[180] At first, much of Japan's foreign policy establishment hoped that the unfolding crisis could be contained, but its resolve steadily weakened as the Kwantung Army advanced across northeast China. When Tokyo discovered a coup plot led by army officers, some civilians in the Japanese government began to believe that they had no choice but to compromise with the army. The government's reluctance to confront the army was reinforced by a rising tide of anti-Chinese sentiment among the Japanese public. Despite their attempts to accommodate the army's plans, the prime minister's cabinet dissolved on December 10, 1931; both the prime minister and cabinet were replaced by those more willing to compromise with the army.[181]

After taking control of northeast China, the Kwantung Army began constructing a puppet state known as Manchukuo without their superiors' permission. Pu Yi, China's last emperor and an ethnic Manchu, would preside over domestic affairs as Manchukuo's head of state. Meanwhile, Japan would control Manchukuo's defense, foreign policy, transportation, and communications. The new Japanese cabinet acquiesced and approved the army's actions in January 1932. Manchukuo declared independence the following month.[182]

---

176 Beasley, 1987, pp. 187–188.

177 Young, 1998, pp. 420–421.

178 Beasley, 1987, p. 174; Young, 1998, pp. 30–35.

179 Thomas W. Burkman, *Japan and the League of Nations: Empire and World Order, 1914–1938*, University of Hawaii Press, 2007, p. 167.

180 Burkman, 2007, p. 167.

181 Lionel P. Fatton, "The Manchurian Crisis as an Exogenous Shock, 1931–1933," in *Japan's Rush to the Pacific War: The Institutional Roots of Overbalancing*, Palgrave Macmillan, 2023, pp. 144–145.

182 Beasley, 1987, pp. 191–194.

Regional and Great Power Reactions

China's strategy for thwarting the Japanese takeover of northeast China focused on appealing to the international community to restrain Japan instead of trying to defeat Japan directly with military force. Japan tried to negotiate bilaterally with China to minimize the League of Nations' involvement.[183]

China's first course of action was to request a meeting of the League of Nations Council. The league responded by passing a resolution that implored both China and Japan to deescalate the situation and requested that both sides withdraw their troops. The League of Nations then called on London and Washington to assist. However, Britain was preoccupied by a domestic economy in crisis, and the United States, which was not technically part of the league, was also hesitant to act. U.S. Secretary of State Henry Stimson alleged that the league was "trying to pass the buck to us" by asking for U.S. assistance.[184]

China then appealed to the United States as a signatory of the Kellogg-Briand Pact, an agreement that ostensibly outlawed war but lacked an enforcement mechanism.[185] The Hoover administration argued that Japan's actions did not violate the pact because the Kwantung Army was acting without Tokyo's approval.[186] Although the U.S. Department of State met extensively with Chinese and Japanese representatives, Stimson feared that any further action would both antagonize Japan and strengthen extremists within the Japanese government.[187]

Following news that the Kwantung Army was bombing Northern China far from the Japanese-controlled railway zone, the United States decided that a proactive response was necessary. Stimson began to believe that Japan's operations were expanding and encouraged President Herbert Hoover to dispatch the U.S. consul general in Geneva to the League of Nations Council.[188] As a result of conversations with the U.S. consul general, the League of Nations Council agreed to remind Japan and China that the Kellogg-Briand Pact obliged them not to resort to war. Shortly thereafter, the council passed another resolution calling for Japan to vacate Chinese territory by November 1931.[189]

Nevertheless, Stimson and the U.S. Department of State hesitated to endorse the league's actions and risk the United States' further involvement. Following the first resolution, Stimson instructed the U.S. consul to remain at the league council as an observer only. Stimson declined to formally endorse the second league resolution; instead, he waited two weeks before sending a watered-down note to Japan that did not mention a deadline for the withdrawal of Japanese troops.[190]

---

[183] Burkman, 2007, p. 167.

[184] Michael E. Chapman, "Fidgeting over Foreign Policy: Henry L. Stimson and the Shenyang Incident, 1931," *Diplomatic History*, Vol. 37, No. 4, September 2013.

[185] Office of the Historian, Foreign Service Institute, "The Kellogg-Briand Pact, 1928," webpage, U.S. Department of State, undated-b.

[186] Richard N. Current, "The Stimson Doctrine and the Hoover Doctrine," *American Historical Review*, Vol. 59, No. 3, April 1954, pp. 513–515.

[187] Chapman, 2013; Current, 1954, pp. 513–515.

[188] Current, 1954, p. 515.

[189] Current, 1954, pp. 516–519.

[190] Current, 1954, pp. 516–519.

Early in the crisis, a core assumption underlying the United States' approach was that going too hard on Japan would only weaken moderates in the Japanese Ministry of Foreign Affairs while strengthening extremists in the Japanese military.[191] But Stimson became convinced that this assumption was wrong by November 19.[192] On December 11, Stimson learned that the Japanese cabinet had dissolved and been replaced.[193]

In January 1932, the U.S. government officials considered working with the league to put economic pressure on Japan. Ultimately, they decided not to because they believed that economic or military pressure would risk inciting a war. Instead, the United States sent a unilateral démarche to Japan and China, which proclaimed that the United States would not recognize any agreement between those two countries that infringed on U.S. treaty rights, the Open Door policy, or the Kellogg-Briand Pact. In short, the United States declined to recognize Japan's annexation of northeast China as legitimate. Stimson called on the other signatories of the Nine-Power Treaty to make similar statements in the hopes of resolving the crisis diplomatically.[194]

The crisis came closest to attracting involvement by the great powers in January 1932, when the Japanese army attacked a Chinese garrison in Shanghai following months of tension between Japanese expatriates and locals. However, this outcome was avoided thanks to a conciliatory Japanese response. Unlike northeast China, Shanghai was home to tens of thousands of foreigners, most of whom were from the United States, Britain, and France.[195] Some previously sympathetic stakeholders began to worry that Japan's actions threatened regional stability.[196]

U.S. government officials dispatched the U.S. Asiatic Fleet to Shanghai on January 31 in response to the Japanese attack on Shanghai. However, officials disagreed on the purpose of this maneuver. Stimson hoped that dispatching the U.S. navy to Shanghai would signal that the United States would defend its interests by any means necessary. In contrast, Hoover was focused on protecting American lives and countered Stimson's intentions by reassuring Japan that the United States did not intend to fight.[197]

Both civilian and military Japanese officials understood that the fighting in Shanghai could attract the ire of the great powers, and they feared that a crisis would endanger the empire's gains in northeast China. Tokyo quickly asked the United States, Britain, and France for help negotiating an end to hostilities with Chinese troops in Shanghai and made sure to consult with those countries on how to protect civilians in the city. Additionally, the commander of Japanese forces in Shanghai avoided hostilities with the U.S. Asiatic Fleet when it arrived.[198] Japan's strategy paid off. On May 5, 1932,

---

[191] Current, 1954, pp. 516–519.

[192] Current, 1954, p. 519.

[193] Current, 1954, p. 519.

[194] Current, 1954, pp. 524–525; Burkman, 2007, p. 167.

[195] Fatton, 2023, p. 147.

[196] Burkman, 2007, p. 169.

[197] Current, 1954, pp. 524–525.

[198] Fatton, 2023, pp. 147–148.

Japan and China agreed to a ceasefire plan for Shanghai that had been devised by the United States, Britain, and France.[199]

Meanwhile, in February 1932, the League of Nations dispatched a commission of inquiry, the Commission of Enquiry, to visit China and Japan. By September 1932, the commission released a report that was relatively balanced. The report recognized Japan's special interests in Manchuria but argued that Japan had not acted in self-defense and had instigated the creation of Manchukuo. Moreover, it recommended that Manchuria be administered in a fashion consistent with "the sovereignty of China."[200]

In December 1932, the League Assembly convened to consider the Lytton Commission's report and created a mediation committee to make recommendations. The mediation committee's recommendations ultimately mirrored those of the Lytton Commission.[201] Japan had hoped that the issue would be resolved in the League Commission because the assembly included some smaller states that sympathized with China.

On February 24, 1933, almost a year and a half after the crisis began, the League Assembly voted 42 to one to accept the mediation committee's report. Only Japan voted against the measure. After completing the vote, the Japanese delegation walked out of the assembly hall—the Japanese Cabinet had voted to withdraw from the league shortly before the vote.[202] Despite this report, on May 31, 1933, the Guomindang signed a truce in which Manchukuo was recognized as an independent state.[203] The following year, Japan abrogated the Washington Naval Treaty that had limited the size of its navy.[204]

Tokyo left the crisis with a new view of the international order and Japan's place in it. Some Japanese policymakers had long believed that the post–World War I international order was biased against Japan. But others—especially in the Ministry of Foreign Affairs—argued that Japan could achieve its goals by working within the system. That faction found itself thoroughly discredited following the great powers' reaction to Japan's invasion of northeast China. In its place rose a new faction—affiliated mainly with the military—who argued that Japan's best course of action was to create a new order.

Aftermath

By mid-1933, the great powers lost interest in the Manchurian crisis. The United States, Britain, France, and the Soviet Union were preoccupied by a combination of domestic issues resulting from the Great Depression and increasingly concerning developments in Europe that would lead into World War II.[205]

---

199 Fatton, 2023, pp. 147–148.

200 Burkman, 2007, p. 170.

201 Burkman, 2007, p. 171.

202 Burkman, 2007, pp. 173–174.

203 Fatton, 2023, p. 153.

204 Ward, 2017, pp. 126–127.

205 Fatton, 2023, pp. 154–156.

Tokyo left the crisis with a new understanding of the international order and Japan's place in it. Some Japanese policymakers believed that the post–World War I international order was biased against Japan, while others—especially in the Ministry of Foreign Affairs—argued that Japan could achieve its goals by working within the system. This second faction found itself thoroughly discredited following the great powers' reaction to Japan's invasion of northeast China. In this faction's place rose a new faction—affiliated mainly with the military—that argued that Japan's best course of action was to create a new order.

Australia remained concerned about Japan's ambitions in the Pacific and began to pursue a more conciliatory foreign policy toward Japan. The Manchurian crisis confirmed the fears of some Australian policymakers that the League of Nations could not provide security in the Asia–Pacific and that, therefore, Australia would have to chart its own course. In 1934, Australia undertook its first diplomatic mission outside the British Empire. External Affairs Minister John Latham visited Japan and tried, and failed, to convince Tokyo to remain in the League of Nations. Latham's efforts at economic diplomacy were more successful: Australia appointed its first trade commissioner to Japan in 1935. However, Australia could not convince the rest of the British Empire to adopt its approach. In meetings with other British Empire leaders in 1935 and 1937, Australian Prime Minister Joseph Lyons proposed a Pacific security pact under which the British Empire would recognize Manchukuo.[206]

Japan was emboldened by the failure of the great powers to act in concert to halt Japan's actions in Manchuria, followed by regional states' accommodations to Japan because of the belief that the great powers would not intervene on the behalf of regional states. Japan continued to pursue its ambitions after the episodes in Manchuria and embarked on the creation of a region-wide Greater East Asia Co-Prosperity Sphere, which encompassed most of East Asia and Southeast Asia.[207] The great powers began to act to turn back Japan only when Japan attempted to extend this empire to U.S. territory.

Key Factors

The key factors that affected the course of events were as follows:

- *Loss of incentive to cooperate:* Although the great powers competed for influence in China during the early 20th century, they also cooperated to maintain the treaty port system. However, the incentive to cooperate dissipated as France, Germany, and Russia's interest in China waned and as the treaty port system crumbled.
- *Loss of stake in the international order:* Although some Japanese policymakers believed that the post–World War I international order was biased against Japan, many nevertheless believed that their country could prosper by working within that order. However, following the collapse of the treaty port system and the league's refusal to accept Japanese rule over northeast China, many in Tokyo came to believe that the great powers would never treat an

[206] Honae Cuffe, "The Manchurian Crisis and the Genesis of Australian Foreign Policy," *The Strategist*, Australian Strategic Policy Institute, January 21, 2022.

[207] Jeremy A. Yellen, *The Greater East Asia Co-Prosperity Sphere: When Total Empire Met Total War*, Cornell University Press, 2019.

Asian nation as an equal and that Japan's best course of action was to spearhead a new international order.

- *Internal dissonance:* Leadership in Tokyo initially did not approve the actions of the local Japanese garrison, the Kwantung Army, when it invaded Manchuria and advanced in northeast China. Once the invasion was underway, the Japanese government belatedly endorsed the occupation and established a puppet state.
- *Distracted international observers:* Other nations were reluctant to engage in the conflict because of the precarious leadership situation in China, Japanese aggression, and a desire to avoid war. This situation worsened as Western nations dealt with the Great Depression and as Europe inched closer to World War II.

Conclusion

By mid-1933, the great powers had lost interest in the Manchurian crisis. The United States, Britain, France, and the Soviet Union were preoccupied by a combination of domestic issues during the Great Depression and increasingly concerning developments in Europe.[208]

The Japanese invasion of Manchuria in the early 20th century arose out of a series of conditions that are likely to arise again. The league's rhetorical condemnations convinced Japan that it would never be considered an equal power in existing international organizations so Japan needed to pursue its own agenda. However, as the Great Depression began, Japan sought to directly control portions of northeast China for economic reasons. Chinese appeals to the international community were met with delayed and disjointed responses from the great powers, which were themselves preoccupied with economic concerns. The lack of international enforcement of the league's previous condemnations convinced Japan that it could continue pursuing an aggressive strategy in Asia without fear of intervention.[209] Similar condemnation is likely to alienate a nation whose territorial objectives clash with the expectations of international institutions. In such a scenario, if the members of the international organization fail to act, the power of the institution is eroded. As a result, similar territorial revisions have the potential to become more frequent as nations are emboldened by the lack of consequences.

---

[208] Fatton, 2023, pp. 154–156.

[209] Ward, 2017, pp. 100–129.

Chapter 5

The Gulf War, 1990–1991

On August 2, 1990, the military of Iraq invaded Kuwait with roughly 100,000 soldiers and significant armor.[210] The Kuwaiti armed forces resisted, but they were largely caught by surprise. In just two days, more than 400 Kuwaiti military personnel were killed and nearly one-half of the Kuwaiti military was captured. By August 4, the Kuwaiti military was defeated, and Iraqi forces occupied the country for the next seven months. Iraq then ruled Kuwait as the country's 19th province, referring to irredentist claims that went back to 1932 and the very founding of the Kingdom of Iraq.[211] Even prior to international intervention led by the United States, the invasion upset the regional alignment in the Middle East. An international coalition force was marshalled, with the United States as the head of the military response to Iraq's aggression.

At the time, Iraq was the head of a burgeoning regional bloc of states, many of which were economically reliant on it. Many around the world saw the Iraqi military as a world-class force—the fifth largest in the world at the time—and there was uncertainty about what the outcome of any ensuing clash would be. However, in contrast to the other cases, the Gulf War is a clear example of a successful rollback of territorial takeover. In particular, Saudi Arabia worried about further Iraqi takeovers: Its main oil fields are about 190 miles southeast of Kuwait's southern border. Saudi Arabia provided the essential staging areas and mustering points for coalition forces, despite significant domestic disapproval.[212] Other Arab states split over Iraq's actions, but countries outside the region widely condemned Saddam Hussein's invasion and provided military and economic support. In fact, these international partners provided so much military, economic, and diplomatic support to the U.S. effort that Washington received a short-term "surplus" to its budget.[213] Potential spillover, norms against territorial revision, and moral violations (e.g., human rights abuses by Iraqi troops against Kuwaitis) isolated Saddam's regime and galvanized a counterinvasion coalition.[214] However, similar normative factors eventually turned global public opinion against continued sanctions on Iraq.

The corollaries between this case and potential future conflicts are straightforward. Iraq aggressed against its smaller neighbor, Kuwait, which did not host any U.S. forces at the time, and Iraq then claimed Kuwait as its 19th province. Kuwaiti forces fought back, but they stood little chance against the qualitatively and quantitatively superior Iraqi military. Iraq then made claims about the legitimacy

---

[210] "August 2, 1990: Iraq Invades Kuwait," webpage, BBC, undated.

[211] Richard Muir, "The Iraq–Kuwait Border Dispute: Still a Factor for Instability?" *Asian Affairs*, Vol. 35, No. 2, 2004.

[212] Clive Jones, "Saudi Arabia After the Gulf War: The Internal-External Security Dilemma," *International Relations*, Vol. 12, No. 6, December 1995.

[213] Leonard Silk, "Economic Scene; The Broad Impact of the Gulf War," *New York Times*, August 16, 1991.

[214] Amnesty International, "Iraq/Occupied Kuwait: Human Rights Violations Since 2 August," December 19, 1990.

of a unification with Kuwait and its rightful ownership of the territory. The lessons included here about the key factors determining regional alignment are applicable to other possible cases of aggression against states that are not official allies of the United States but in which the United States has significant interest.[215]

Historical Context

Regional Alignment

Prior to the conflict, the Arab states of North Africa and the Middle East were divided into three key blocs. First was the Arab Cooperation Council (ACC), consisting of Egypt, Iraq, Jordan, and North Yemen. Second was the Arab Maghreb Union, which included the five North African states west of Egypt. Last was the Gulf Cooperation Council (GCC), made up of the Gulf Arab states, including Kuwait, with the exclusion of Iraq.[216] These three blocs were seen as the main dividing lines in the region and the likely way in which the region would split in the event of any kind of crisis.

In early 1990, tensions between Egypt and Iraq on visions of the future of the region began to rise. Much of this was because of ongoing economic and political difficulties in the USSR, arising partly because of the Soviet Union's failed war in Afghanistan, which led to an absence of the Soviet Union in the region as the Cold War ended. With the loss of Russia's deep involvement in the Middle East and the United States's regional dominance, Arab states had to face important questions as to how they should respond.[217]

Without the balancing power of the Soviet Union in the region, there was a fear that countries would be exploited by the West. This argument was taken up by Prime Minister Saddam and other Iraqi officials, who looked to tie Arab states more tightly together in a unified bloc to counter Western influence.[218] Iraq's easing of relations with the United States during the Iran-Iraq War (1980–1988) was always tentative and done only when the Soviet Union refused to resupply Iraqi forces after initial campaigns.[219] This reopening was a drastic shift from the norm in which Iraq and the United States had no diplomatic relations for the previous 16 years. After the conclusion of the war—and in the absence of the United States' need to compete with the Soviets for Iraqi favor—the Iraqi government's concern that the United States would exploit Iraq returned. Iraq argued that the ascension of the United States to the top of the global system would mean that Israel would be emboldened to undertake adventurism beyond its existing borders.[220] Iraq further argued that Gulf countries should

[215] For the purposes of our analysis, we confine the history primarily to the period immediately prior to the invasion and the period prior to Operation Desert Storm.

[216] Iraq had been an associate member of the GCC prior to 1990. Bruce Maddy-Weitzman, *The Inter-Arab System and the Gulf War: Continuity and Change*, Vol. 1, Carter Center of Emory University, November 1991.

[217] George Joffe, "Middle Eastern Views of the Gulf Conflict and Its Aftermath," *Review of International Studies*, Vol. 19, No. 2, April 1993; Maddy-Weitzman, 1991.

[218] Peter W. Rodman, "Middle East Diplomacy After the Gulf War," *Foreign Affairs*, Vol. 70, No. 2, Spring 1990, p. 1.

[219] Joffe, 1993.

[220] Maddy-Weitzman, 1991.

remove any U.S. military presence. It asserted that U.S. ascendance threatened the sovereignty and security of countries in the Middle East, especially in any conflict with Israel in which the United States would inevitably back Israel and use U.S. military forces against countries that opposed Israeli adventurism.[221]

In contrast, Egypt argued that the more appropriate response was to integrate further with the West, particularly with European states that could check any potential U.S. excesses.[222] Egypt opposed the aggressive rhetoric from Iraq and saw it as generally unhelpful and imperiling potential investment from the West. Additionally, arms infusions from the United States played a large role in Egyptian politics, and Egypt felt that distancing itself from the United States would make a conflict between regional actors and Israel or the United States likelier. The Egyptian government accused Jordan of being an Iraqi puppet, adding ACC agenda items that suited Iraq's interests and promoting the perception of the ACC abroad as an anti-American bloc.[223] Unlike Iraq, which could rely on oil revenue, Egypt lacked similar resources and was reliant on trade with and aid from external actors. Thus, Egypt was upset that Iraq and Jordan were signaling to the United States and the wider West that the ACC was universally opposed to U.S. and European connections. Because Egypt was simultaneously trying to strengthen those ties, such actions were contrary to Egypt's interests.

This regional split prior to the Iraqi invasion of Kuwait culminated in the Arab League's 1990 Baghdad Summit in which Iraq stated its position in opposition to Western influence in the region and affirmed its right to maintain weapons of mass destruction as protection against the aggressive West.[224] Such a statement indicated that the region's overall alignment just prior to the invasion was a clear split between those states that were predisposed to having friendly relations with the West and those that sought to keep the Arab world free from Western influence and as a distinct power center of its own.

Iraq led the camp intent on establishing a regional bloc to counter Western influence, with Jordan, north Sudan, and some North African states, such as Libya, acquiescing. This alignment split the ACC, with Egypt largely against the rest of the regional bloc. It also split the Arab Maghreb Union, with Algeria and Libya opposing Western influence and Morocco being more open to it. Last, the GCC was split as well, with Saudi Arabia largely opposed to having U.S. military forces in Gulf countries and criticizing the United Arab Emirates and Bahrain for maintaining a U.S. military presence. Other smaller states and groups followed one of these larger blocs; for example, the Palestine Liberation Organization (PLO) followed Iraq's lead and received significant funding from Saddam's regime.[225] Iraq and other states with large resource endowments felt freer to set a course of opposition to Western influence because they were less reliant on other forms of trade with the West or aid from Western sources.

---

[221] Maddy-Weitzman, 1991.

[222] Joffe, 1993; Maddy-Weitzman, 1991.

[223] Maddy-Weitzman, 1991.

[224] Maddy-Weitzman, 1991.

[225] Maddy-Weitzman, 1991.

Great Power Alignment

Prior to the invasion of Kuwait, the great powers were largely hands off toward the Middle East. The Soviet Union had retreated from many of its commitments prior to 1990, partially because of a lack of resources as the Soviet economy suffered. In addition, the Soviets stated that they no longer saw a connection between the Gulf and the overall East-West military balance.[226] These Soviet activities spurred the debate between regional states on how the Middle East should arrange itself in the absence of a second great power.[227]

Without concern for Soviet influence, the United States was less inclined to deeply involve itself in Middle Eastern affairs outside some key relationships with Israel, Egypt, and Saudi Arabia. The United States' avoidance of involvement in the Middle East was evident prior to the Iraqi invasion. When prompted by Iraqi officials on the U.S. stance regarding the country's disagreements with Kuwait, U.S. ambassador to Iraq April Glaspie stated that the U.S. government had no opinion about the issue or Arab-Arab conflicts.[228] The ambassador also stated that the United States would not start an economic war against Iraq. When the Assistant Secretary of State testified before Congress and emphasized that the United States had no treaty obligations to Kuwait. Rather, the United States simply declared its concern in the event that Iraq crossed the border.[229]

Of the European states, France sought to take on the role of interlocutor between the Arab world and the West. This meant dealing with states, such as Egypt, that sought closer relations with Europe, particularly on trade. It also took the form of a close relationship with Iraq by engaging in significant trade, including supplying weapons and selling nuclear reactors to Iraq. France hoped to use these positive relations across regional divides to integrate the European community.[230]

Overall, none of this indicated that there would be a massive diplomatic and military effort in the event of a conflict. The United States overtly signaled that a conflict between Iraq and Kuwait would not require a significant response, and the Soviets had largely abandoned the region as an area of interest.[231] Because France aimed to have positive relations across all the different regional blocs, there was also little sign that such a move from Iraq would draw harsh condemnation from French leaders.

Takeover

On August 2, 1990, Iraq invaded Kuwait; it was immediately clear that Kuwaiti opposition was unlikely to last long. In the wake of the invasion, the existing regional schisms defined state responses.

---

[226] Benjamin S. Lambeth, *Desert Storm and Its Meaning: The View from Moscow*, RAND Corporation, R-4164-AF, 1992.

[227] Rodman, 1991; Fred Halliday, "The Gulf War and Its Aftermath: First Reflections," *International Affairs*, Vol. 67, No.2, 1991.

[228] "Confrontation in the Gulf; Excerpts from Iraqi Document on Meeting with U.S. Envoy," *New York Times*, September 23, 1990.

[229] John M. Goshko and R. Jeffrey Smith, "State Dept. Assailed on Iraq Policy," *Washington Post*, September 19, 1990; Elaine Sciolino and Michael R. Gordon, "Confrontation in the Gulf; U.S. Gave Iraq Little Reason Not to Mount Kuwait Assault," *New York Times*, September 23, 1990.

[230] Pia Christina Wood, "François Mitterrand and the Persian Gulf War: The Search for Influence," *French Politics and Society*, Vol. 10, No. 3, Summer 1992.

[231] Sciolino and Gordon, 1990.

Some states lined up behind Iraq in support of its actions. Others expressed immediate opposition to the idea of Western forces intervening in an intra-Arab conflict. Despite these differences, the UNSC swiftly condemned the invasion and began to lay the legal foundation of an international response.

Regional Reactions

Most of the ACC and Arab Maghreb Union expressed support for Iraqi actions during the invasion and in the hours following. This support included the open alignment of the PLO, North Yemen, Jordan, Sudan, and Mauritania with Iraq, along with statements from Libya and Tunisia that were slightly more ambiguous but had little hint of disapproval of Iraq's actions.[232] Of all the Arab states, only Morocco and Algeria issued immediate condemnations of the invasion and occupation, although Algeria would play both sides of the issue, harming its relations with both the pro- and anti-Iraq blocs.[233]

Although the region had divided views of the invasion, there was agreement in opposing Western intervention. The Arab League criticized the UNSC's quick condemnation, claiming that the events were an Arab issue to be dealt with solely by the league. After the UNSC's action to condemn the invasion, demand a withdrawal, and put all options on the table if Iraq did not withdraw, many in the region thought that such a condemnation was a pretext for Western intervention, and the Arab League quickly made its opposition to Western involvement clear. Members traded bitter accusations among themselves over who was to blame for the introduction of Western troops into the Middle East even prior to the crisis and how much responsibility the West had for the invasion itself.[234]

The biggest divergence from preexisting regional splits came from Saudi Arabia. When Saddam began rhetorically threatening the Saudi regime after the initial invasion of Kuwait, the regime's core legitimacy was undermined. Saudi Arabia's main oil fields in the northeast of the country were under direct threat. When Iraqi forces made incursions into northeast Saudi territory, the Saudi regime saw a nexus of events that required it to take drastic action that diverged from previous policies because its political legitimacy was under threat from multiple angles.[235]

The Saudi government knew that it needed assistance to meet the threat from Iraq, but the source of that assistance was unclear. Arab militants offered support, but Saudi Arabia had undertaken a deliberate policy to export its domestic militancy during the 1980s.[236] The idea of inviting Arab militants back into Saudi territory was seen as courting disaster as it was believed that these extremists would likely undermine the regime in the aftermath of a conflict. The second option was to accept support from Arab nationalist forces—specifically, Egypt and Syria. However, Arab nationalism was seen as a threat to the Saudi regime as well because a victory for these outside forces could again

---

[232] Joffe, 1993.

[233] Joffe, 1993.

[234] Ziad Swaidan and Mihai Nica, "The 1991 Gulf War and Jordan's Economy," *Middle East Review of International Affairs*, Vol. 6, No. 2, June 2002, p. 71; Maddy-Weitzman, 1991.

[235] Jones, 1995.

[236] Jones, 1995.

undermine the regime and show prowess in the face of Saudi weakness.[237] This threat was not overblown: In 1992, the Islamist victory in Algerian elections was quickly overturned by a military coup.[238]

The remaining option was to extend an invitation to the United States and the U.S.-led coalition forces. From the Saudi perspective, this was the least objectionable of a set of bad options. The United States was the least likely to undermine Saudi legitimacy directly, was the likeliest to win in the conflict, and was also the likeliest to rein in other coalition partners, particularly the Egyptians and Syrians. The United States promised that its forces would go home after the conflict and that it would provide additional weapons to the Saudi regime in the future. Because of these terms, the Saudis believed that U.S. help was the best option to protect the regime both during the crisis and after.[239]

Although this was the best path the Saudis saw for themselves at the time, it was not without regional repercussions. In numerous countries, there were significant protests against the presence of non-Muslim troops on the Arabian Peninsula. Saudi Arabia, Egypt, Morocco, and Tunisia all experienced large protests, but they had minimal results. Even larger protests played a role in Algerian policy. Algeria initially condemned the invasion, but the outpouring of disapproval for the presence of Western forces convinced the government to change positions and assert that what was occurring was U.S. aggression and meddling in regional affairs.[240]

Alternatively, Libya's Muammar Gaddafi initially took an anti-U.S. and anti-British stance in keeping with his traditional animosity toward the West. However, he was convinced to change course by Egyptian President Hosni Mubarak, who warned that Western ire may turn to Libya after the crisis subsided.[241] As a result, Libya's policy afterward was cautiously pro-Western to avoid sticking its neck out in the face of a set of triumphant Western powers.

Great Power Reactions

The invasion was unanimously condemned by world powers. Condemnations flowed in from France, India, China, the Soviet Union, the UK, and the United States. Iraq's main sources of weapons, China and Russia, also put an arms embargo in place.[242] The unanimity across world powers set the stage for immediate UNSC resolutions condemning the invasion and calling for an international response, including sanctions.[243] This opposition was the result of fears regarding steady

---

[237] Josh Pollack, "Saudi Arabia and the United States, 1931–2002," *Middle East Review of International Affairs*, Vol. 6, No. 3, September 2002.

[238] Joffe, 1993.

[239] Pollack, 2002; Louisa Dris-Aït-Hamadouche and Yahia H. Zoubir, "The US-Saudi Relationship and the Iraq War: The Dialectics of a Dependent Alliance," *Journal of Third World Studies*, Vol. 24, No. 1, Spring 2007.

[240] Joffe, 1993.

[241] Joffe, 1993, pp. 177–199.

[242] Bill Keller, "The Iraqi Invasion; Moscow Joins U.S. in Criticizing Iraq," *New York Times*, August 4, 1990; David Holley, "China to Join Iraq Arms Embargo," *Los Angeles Times*, August 6, 1990; Lambeth, 1992.

[243] UNSC Resolution 660 and later Resolution 678 demanded withdrawal and allowed "all necessary means" by responding states to force Iraq out of Kuwait (UNSC Resolution 660, adopted by the Security Council at its 2,932nd meeting, August 2, 1990; UNSC Resolution 678, adopted by the Security Council at its 2,963rd meeting, November 29, 1990.

oil supplies, stories of Iraqi human rights abuses in Kuwait, and the need to enforce global norms against territorial aggression as the Cold War wound down.

The events changed the views of U.S. political officials regarding the region. The invasion presented a serious threat to global norms against territorial revision that decisionmakers saw as needing enforcement. UK Prime Minister Margaret Thatcher reminded U.S. President George H. W. Bush of the potential costs of appeasement, hearkening back to the deals made by Western powers in the 1930s with Adolf Hitler.[244] In addition, the possibility of an overtly anti-Western figure controlling most of the world's oil supply was seen as a severe threat to the global economy, particularly to highly developed Western economies that relied on a steady supply of cheap Gulf oil.[245] This combination of issues raised the stakes of simply accepting a fait accompli in Kuwait. If the invasion were allowed to stand, Saddam would continue to gain in power and leverage over time, with further threats likely to come. Reports of widespread human rights abuses by Iraqi forces in Kuwait further entrenched the image of Iraq as an aggressive power bent on domination and unbeholden to global norms.

These reactions arose from a nearly universal recognition that the invasion was a violation of the global norms that countries wished to live by in the aftermath of the Cold War and was a direct threat to the global economy if Iraq held leverage over a substantial portion of the world's oil supplies. With the Soviet Union weakened, countries recognized that the United States was emerging as the unrivaled global power. In the absence of global norms protecting the sovereignty of smaller states, weaker states throughout the world could be under threat from their larger and more-powerful neighbors. The Iraq invasion presented an opportunity to strengthen global norms to protect territorial integrity and the rights of smaller and weaker states.[246]

The USSR's opposition to Iraq's invasion surprised many, given the long-standing relationship between the two countries and the hefty supplies of arms that the USSR provided Iraq over time. However, the Soviet government calculated that supporting Iraq would be "counter to the new political thinking and civilized relations between states" that the USSR was trying to foster with the noncommunist world.[247] There was a hope in Moscow that cooperation on Iraq would stand as an example for other areas of emerging U.S.-Soviet cooperation, including economic support for the crumbling Soviet regime.[248] This cooperation went so far as an offer to provide Soviet troops to join the U.S. coalition; the request was denied because the Desert Shield operation had already shown success without it.[249] Although there were some domestic actors within the Soviet Union (particularly in the military) that still argued for resupplying Iraq, these voices were sidelined in favor of a

---

[244] Tom Shachtman, "It's Time to Abandon 'Munich,'" *Foreign Policy*, September 29, 2013.

[245] Margaret Thatcher Foundation, "Saddam Invades Kuwait (24 Jul–2 Aug 1990)," webpage, undated.

[246] Jean-Marc Rickli and Khalid Almezaini, "Theories of Small States' Foreign and Security Policies and the Gulf States," in Khalid Almezaini and Jean-Marc Rickli, eds., *The Small Gulf States: Foreign and Security Policies Before and After the Arab Spring*, Routledge, 2016; Elizabeth Riddell-Dixon, "The United Nations After the Gulf War," *International Journal*, Vol. 49, No. 2, 1994.

[247] Lambeth, 1992, p. 8.

[248] Lambeth, 1992.

[249] Lambeth, 1992.

cooperative stance.[250] Outside analysts immediately saw that, in the absence of Soviet resupply, Iraq would not be able to withstand a coalition assault.[251]

China also sided with the coalition to more firmly establish itself as a responsible great power actor in global affairs and to set the stage for greater cooperation on economic issues, such as trade—both of which were paramount for the Chinese regime. Despite the overt cooperation, there were still efforts to test its limits. For instance, China pondered covert trade deals with Iraq to give China preferential treatment for Iraq's oil in the aftermath of the conflict.[252] This issue showed the paramount importance of economic factors for China, but the overall benefits of firmly positioning itself as a normal, responsible state to produce trade benefits around the world won the day in Beijing. In addition, Beijing made the calculation that stable access to Gulf oil into the future was paramount and superseded any gains it could make through preferential oil arrangements with Iraq itself.[253]

France and the UK saw the issue as a way to bolster their global roles. France attempted to use the conflict to enhance its position as the mediator of choice in the Middle East by linking the Kuwait situation with an Israel-Palestine peace agreement.[254] This idea was dismissed out of hand by the United States, which opposed any sort of linkage. The UK also saw involvement in the coalition effort as an opportunity to advance arms sales to the wealthy Gulf states.[255]

Overall, great power opposition to Iraq's invasion arose from a confluence of events. In the waning days of the Cold War, states sought ways to ensure their rights so that smaller and weaker actors would be protected against an empowered United States. With the United States playing the role of the guarantor of territorial integrity and sovereignty, these countries saw a way to move the United States into a position of supporting them. By working through the UN, the United States also established the UN as a further check that countries could use to tie the hands of the unipolar power.[256]

Different countries pursued their own domestic political goals as well. UK and U.S. ambitions to bolster their domestic arms industries played a role in the decision to intervene and boost their domestic economies.[257] Stories of Iraqi's human rights abuses in Kuwait also established a straightforward narrative of Iraq as the villain in Western political discussion.[258] Such stories demanded action from Western powers and lowered the political capital that leaders would have to expend for intervention. The Soviets were consumed with their own domestic problems, which prevented them from embarking on any meaningful foreign activity, so Moscow sought to cooperate with the West. The threat to Saudi oil catalyzed significant international opposition because a shock

---

250 Alvin Z. Rubinstein, "Moscow and the Gulf War: Decisions and Consequences," *International Journal*, Vol. 49, No. 2, 1994.

251 Lambeth, 1992.

252 Rowland Evans and Robert Novak, "An Erosion of Support," *New York Times*, September 20, 1990.

253 John W. Garver, "China and American Hegemony in the Persian Gulf," in *China's Quest: The History of the Foreign Relations of the People's Republic of China*, Oxford University Press, 2018.

254 Wood, 1992.

255 Teemu Hakkinen, "Parliament and the Gulf War in 1991," *History of Parliament* blog, January 20, 2016.

256 Ian Hurd, *After Anarchy: Legitimacy and Power in the United Nations Security Council*, Princeton University Press, 2008.

257 Hakkinen, 2016; Pollack, 2002.

258 Philip Alston, "The Security Council and Human Rights: Lessons to Be Learned from the Iraq-Kuwait Crisis and Its Aftermath," *Australian Year Book of International Law*, Vol. 13, 1992, p. 107.

to oil markets—and its resulting economic disruptions—would imperil nearly every leader's political survival.[259]

Aftermath

After the coalition effort pushed Iraqi military forces out of Kuwait, many saw the outcome as an unmitigated triumph for the United States. Although this was true in a military sense, and in the immediate aftermath of Desert Storm, adaptations in Iraq's public relations strategy meant that this sense of victory did not last. The aftermath saw continued engagement from the United States, with the imposition of severe sanctions against Iraq and the patrolling of restrictive no-fly zones in both Iraq's north and south by U.S. military forces.

Regional Alignment

Immediately after the conflict, the regional alignment shifted dramatically against Iraq and in favor of the United States. Many of the countries and organizations that had supported Iraq's invasion apologized or changed course, realizing they had backed the losing side. Part of these changes were the result of tangible actions by Kuwait in the aftermath of the conflict, such as the deportation of nearly 400,000 Palestinians, who were fired from government jobs and expelled from educational institutions.[260] Similar results occurred in North Yemen, which saw a collapse in aid from the United States, Saudi Arabia, and Kuwait.[261] Last, Jordan's alliance with Iraq dissolved. With Iraq no longer able to provide sufficient economic resources to Jordan because of sanctions, Jordan made peace with Israel and began a gradual alignment with the United States.[262] In total, much of the region abandoned Iraq and suffered consequences at the hands of the victorious coalition, which prompted immediate rhetorical shifts and policy changes that isolated Iraq and minimized open opposition to the United States.[263]

However, two developments later shifted alignments. First, sanctions against Iraq began to take their toll on the Iraqi economy and population. Second, Iraq shifted its rhetoric from one of a strong power standing up to the imperialist West to one of a weak and helpless state being bullied by a global superpower. These changes altered regional alignments over time.[264]

---

[259] Robert J. Lieber, "Oil and Power After the Gulf War," *International Security*, Vol. 17, No. 1, 1992; Paul J. Burke, "Economic Growth and Political Survival," *B. E. Journal of Macroeconomics*, Vol. 12, No. 1, 2012.

[260] Joffe, 1993.

[261] Sheila Carapico, "Yemen: Unification and the Gulf War," *Middle East Research and Information Project*, Vol. 170, May–June 1991.

[262] Curtis Ryan, "Between Iraq and a Hard Place: Jordanian-Iraqi Relations," *Middle East Report*, Vol. 215, Summer 2000, pp. 40–42.

[263] Yahya Sadowski, "Arab Economics After the Gulf War," *Middle East Research and Information Project*, Vol. 170, May–June 1991.

[264] Samuel Helfont, "The Gulf War's Afterlife: Dilemmas, Missed Opportunities, and the Post–Cold War Order Undone," *Texas National Security Review*, Vol. 4, No. 2, Spring 2021.

Iraq began putting out information that Western sanctions were leading to the death of tens of thousands of children in a cruel collective punishment. Studies of the Iraqi economy were highly publicized by the regime in an effort to play to the sympathies of Arab publics and a wider global audience. These efforts lent credibility to the Iraqi regime's arguments that it was now the target of inhumane treatment by an imperialist bully; over time, these arguments had a significant effect on regional alignment.[265]

By changing the rhetorical narrative, Iraq shifted the discussion of global norms from one in which the United States was upholding global norms to one in which the United States was the key actor undermining them, failing to respect sovereignty, and engaging in human rights abuses. By the time of the September 11, 2001, terrorist attacks against the United States and the run-up to the Iraq War, these efforts had been successful enough to turn much of the region, which refused to support U.S. operations into Iraq, against the United States—even when offered tens of billions of dollars in compensation. Because Iraq's influence efforts had been so successful, even if Middle Eastern regimes supported U.S. policy, it was untenable to act against large and active majorities of their populations who vehemently opposed U.S. action.[266]

Great Power Alignment

Iraq's strategies during this postwar period were highly attuned to local contexts in non-regional actors. For instance, France's participation in the Gulf War was always justified as a way to reestablish regional security and support territorial integrity. The French public and government never accepted the idea of a new world order that was being ushered in by coalition efforts. Once regional security and territorial integrity had been reestablished, the French viewed continuing sanctions, the patrol of no-fly zones, and repeated air strikes on Iraqi assets within these no-fly zones as *themselves* undermining regional stability. The French were also susceptible to Iraqi arguments that the Western sanctions were leading to widespread suffering and misery, amounting to an ongoing human rights crisis.[267]

A similar series of developments occurred with Russia. As Russia's humanitarian crisis continued, the Russian government increased its criticism of U.S. policies. Iraqi officials heavily lobbied the Russian government, engaged in influence operations in the Russian press, and exploited Russian divergence from U.S. policy in the Balkans in the mid-1990s. Through the Balkan policies, the expansion of NATO, and these Iraqi efforts, relations between the United States and Russia soured over the next decade and finally led to the withdrawal of the Russian ambassador to the United States over air strikes on Iraq in 1998. The culmination of the degradation in this relationship was the full-throated Russian opposition to the Iraq War in 2002 and 2003.[268]

These strategies were highly successful for Iraq. By the late 1990s and early 2000s, the UN was acting largely in opposition to U.S. policies in Iraq. Iraq had successfully influenced key members of the UNSC, which then acted as roadblocks to U.S. proposals on Iraq. These actors undercut the

---

[265] Helfont, 2021.

[266] Joshua Rovner, "Delusion of Defeat: The United States and Iraq, 1990–1998," *Journal of Strategic Studies*, Vol. 37, No. 4, 2014; Helfont, 2021.

[267] Helfont, 2021.

[268] Helfont, 2021.

legitimacy of U.S. actions by denying the United States approval from the primary legitimacy-granting body on the world stage. UN approval had played a key role in rallying the world against Iraq in 1990 and 1991, but it played the opposite role by the time of the Iraq War. By successfully influencing global public opinion, and therefore government action, Iraq reframed the debate from one in which it was a bully to one in which it was being bullied. Thus, the reasons for global alignment behind the United States during the Gulf War were turned on their heads by the time of the Iraq War.

Key Factors

The key factors that affected the course of events were as follows:

- *Economic linkages:* Many countries that relied on Iraqi resources initially sided with it in the conflict, whereas those countries that were more integrated with Western trade sided with the coalition. Global reliance on energy supplies from the region galvanized support for the coalition, and the possibility of selling arms to wealthy Gulf states further incentivized Western involvement.
- *Threat perceptions:* Countries that perceived an external threat were likelier to balance against it rather than bandwagon with the aggressive power. In addition, fears of domestic populations determined many states' alignment decisions.
- *Global norms and legitimacy:* A change occurred in the international community's view of the United States. First seen as upholding global norms, the United States was now seen as undermining norms because of its economic sanctions and air strikes against Iraq, which were seen as a violation of human rights. This change shifted international public opinion against the United States in its policies on Iraq. Approval through both regional and international bodies, such as the Arab League and the UN, also played a significant role in alignment choices. Last, states seen as violating human rights norms were opposed to U.S. involvement. Initially Iraq was viewed as the violator, but, in the conflict's aftermath, it was the United States.
- *Influence strategies:* Iraq's public relations strategy around the world shifted from one of portraying itself as a strong opponent to the United States to one in which it was a helpless victim against American bullying. Continued sanctions and air strikes seemed to reinforce this positioning of Iraq against the United States. Iraq's strategy also tailored its public relations strategies to individual great powers and their domestic contexts, using views within domestic political contexts that steered policies and opinions away from the United States.

Conclusion

Prior to Iraq's invasion of Kuwait, the Middle East was split into subblocs based on views of what the region should look like in the post–Cold War order. These blocs largely mapped onto how countries responded in the immediate wake of the invasion, with countries that backed Iraq having previously accepted its vision of an independent region that could stand against Western influence. The alternative camp, led by Egypt, sought to enhance relations with the West and integrate the Middle East with the wider world. Only a few countries diverged from these preexisting schisms, with the largest outlier being Saudi Arabia, mainly because of the immediate danger posed by Iraq to its borders and regime legitimacy.

The success of the coalition's operations to liberate Iraq shifted this alignment in favor of the United States and against Iraq. Such countries as Jordan and Yemen, as well as the PLO, were forced to change their stances to account for the ascendance of the United States and a weakened and isolated Iraq. However, this shift was short-lived: Iraq shifted its influence strategies to portray itself as a weak actor being subjected to bullying by a superpower. This strategy was successful, and it eroded support for the United States on the global stage, leading to large portions of Arab publics seeing the United States as a nefarious power causing a humanitarian crisis in Iraq. Such a shift significantly inhibited further U.S. policies directed at Iraq, whether the sanction regime during the 1990s or the lead-up to the Iraq War in the early 2000s.

Spillover was a key factor motivating Saudi Arabia's hosting of coalition forces to repel Saddam's invasion. An Iraqi capture of a significant portion of global oil sources inspired similar concerns outside the region. The United States specifically benefited from tailwinds in support of global norms against territorial aggression and its burgeoning victory in the Cold War.[269]

The key factors here show the importance of influence strategies and public opinion and, particularly, the ability of those factors to constrain action and reduce international legitimacy. The United States went from being viewed as the ultimate enforcer of global norms with widespread international support and legitimacy to a subverter of global norms acting without regard for international views because of the ongoing sanctions and bombing of Iraq. Within this context, more-specific factors played important roles, such as the immediate threat perception that countries had, their dependence on different international economic partners, and views of who would likely be the winner of the conflict.

Potential future conflicts can find significant corollaries in this case. Preexisting regional schisms are likely to stay in place in the absence of new dynamics because they are heavily related to existing economic linkages. Changes can occur when a regional state sees a direct threat to itself after a territory is taken, perceives how the acquired population will be treated in the conflict's aftermath, understands the legitimacy of actions on either the attacker or defender side, and sees how particular influence strategies might sway publics and governments around the world.

---

[269] G. John Ikenberry, *After Victory: Institutions, Strategic Restraint, and the Rebuilding of Order After Major Wars*, Princeton University Press, 2019.

Chapter 6

The Crimean War, 1853–1856

In 1853, Imperial Russia invaded the Romanian principalities, claiming the right to protect Orthodox Christians in the primarily Muslim Ottoman Empire. In response, France and Britain moved fleets to the Black Sea to dissuade Russian advances. In October 1853, the Ottoman Empire declared war on Russia, then on Britain and France in March 1854.[270] Despite easily outmatching Turkish troops, Russian forces failed to elicit support from Prussia or Austria and struggled against the combined militaries of Britain and France. After several messy campaigns surrounding the Black Sea, Sevastopol, the Balkans, and the Caucasus in December 1855, Austria issued an ultimatum to Russia in which it threatened to join the war on the side of the British-French coalition if a peace was not reached. In January 1856, an agreement was reached that compelled Russia to give up protection rights to Turkish Orthodox Christians, neutralized the Black Sea, surrendered southern Bessarabia, and opened the Danube River to international trade. Additionally, new international laws were developed by the coalition that abolished privateering and established laws on neutral flags and rules on trade during conflicts.

The Russian Empire ostensibly invaded to protect coethnics living in Ottoman territory. Moreover, St. Petersburg sought advantage in the Eastern Question (the question of what to do with territories being ceded by the waning Ottoman Empire) against other European powers that were also keen to divide up and gain influence over territories within a declining Ottoman Empire. The possibility of spillover—continued Russian conquest and Russian access to the Mediterranean—prompted London and Paris to support the Ottomans militarily to contain St. Petersburg's gains and allow the Sublime Porte to hold out against further losses. That cooperation between England, France, and the Ottoman Empire was plagued by distrust and conflicting goals, and it collapsed soon after the end of the war.

In this case, three comparisons can be generalized to apply to potential future scenarios. First, Russia acted as the aggressor and justified its invasion of another nation's borders as an attempt to protect people of a similar demographic profile to its own. Second, the Ottoman Empire was much weaker than the Russian forces and quickly proved incapable of standing up to Russian advancement quantitatively or qualitatively. Finally, Russia was such a powerful force at the time that a coalition of the UK and France, fearing further Russian expansion, had to aid the Ottoman Empire to successfully ward off Russia's invasion. The dynamic between these three international players—the aggressor with cultural interests in territorial expansion, the weak nation, and the strong coalition forces that

[270] David M. Goldfrank, *The Origins of the Crimean War*, Routledge, 1994; John Sweetman, *The Crimean War*, Routledge, 2001; Andrew Lambert, *The Crimean War: British Grand Strategy Against Russia, 1853–56*, 2nd ed., Routledge, 1994; Candan Badem, *The Ottoman Crimean War (1853–1856)*, Brill, 2010.

intervene—can easily be applied to contemporary cases with similar power dynamics. This shows a coalition's ability to overcome the interests of a single, powerful state that seeks territory.

Historical Context

In the decades leading up to the Crimean War, it had become clear to most international observers that the Ottoman Empire was in a protracted state of decline that was punctuated by domestic unrest, poor economic prospects, and corruption. The Ottoman Empire covered a large number of territories, peoples, and cultures, and the efforts to hold these differing groups together had led to revolts around the empire.[271] The leader of the Ottoman Empire, Sultan Abdülmecid I, was considered weak and indecisive both on the international stage and domestically, which led Ottoman political and military officials to begin focusing their attention on self-advancement and personal enrichment through bribery, corruption, and theft.[272] This left the Ottoman army in a state of significant disrepair: Food, supplies, and training to carry out an effective operation were all lacking. As a result, other European powers often aided the Ottoman Empire in quelling rebellions and providing aid to avoid the collapse of the Ottoman Empire.[273] European powers had an interest in preventing a scramble for ceded Ottoman territories among rival powers, especially Russia. If Russia were allowed to gain additional territory in the Balkans through Asia Minor and into the Middle East, it could become a naval force in addition to a land power, begin cutting off trade routes further east, and reduce territorial buffers to further Russian expansion.

Regional and Great Power Alignment

Prior to the outbreak of the Crimean War, Russia was concerned primarily about domestic and regional rebellion and sought to solidify imperial power and stabilize the region in the face of increasingly frequent revolutions throughout Europe. Russia's leader, Tsar Nicholas I, was greatly concerned about the rise of rebellions and fall of monarchies that had occurred throughout Europe in the 1830s and 1840s and began a focused campaign to maintain the status quo, earning himself the nickname the "gendarme of Europe."[274] Russia was considered the land superpower of the age because of the size, experience, and strength of the Russian military. However, Nicholas I believed that Russia could work with European leaders to ensure the stability of the region.[275]

---

[271] Badem, 2010.

[272] Badem, 2010.

[273] Leslie Rogne Schumacher, "The Eastern Question as a Europe Question: Viewing the Ascent of 'Europe' Through the Lens of Ottoman Decline," *Journal of European Studies*, Vol. 44, No. 1, 2014.

[274] In these campaigns, Russia would aid in suppressing revolts then quickly withdraw back across Russian borders (Andrei P. Tsygankov, "The Crimean War, 1853–1856," in *Russia and the West from Alexander to Putin: Honor in International Relations*, Cambridge University Press, 2012, p. 204).

[275] These attempts at stability included a war with Iran in 1826 and a war with the Ottoman Empire in 1828. Andrew Lambert, "The Anglo-French Crimean War Coalition, 1854–1856," commentary, Royal United Services Institute for Defence and Security Studies, March 16, 2010; Badem, 2010; Schumacher, 2014; Alexandra T. Evans, *Alternative Futures Following a Great Power War*: Vol. 2, *Supporting Material on Historical Great Power Wars*, RAND Corporation, RR-A591-2, 2023.

A series of contradicting treaties between the Ottoman Empire, Russia, and France gave unclear jurisdiction on the rights of Russia to protect the interests of Ottoman Christians. These documents were ambiguous and easily distorted, which led the tsar to believe that his rights were violated by French declarations of authority over these Christian groups.[276] Nicholas I believed that the legitimacy of Russia was connected with his rights as a Christian power to protect Ottoman Christians, and Russian strategists considered the Balkan region to be highly dangerous and volatile.[277] As it became clear that Nicholas wanted to intervene on behalf of the Ottoman Christians, Russian strategists developed a plan for war with the Ottoman Empire that they hoped would ensure that other nations would not intervene.[278] Russian observers saw the Ottoman Empire as able to engage in only a small war but also noted the Russian navy was not as strong as those of Britain and France; therefore, they sought to contain events to a bilateral dispute between the Ottomans and Russians.[279] In 1844, Nicholas I traveled to Britain to assure British officials that Russia wanted to keep the Ottoman Empire stable and did not want its territory but also noted that he would not allow other European powers to gain hold of it.[280]

The Crimean War is often cited as the first modern war because it was fought primarily over the growing concerns regarding the Eastern Question rather than questions of religion. The Eastern Question encompassed a variety of concerns, such as power vacuums, ethnic representation, and how to divest the assets of the flagging Ottoman Empire. Whereas the Vienna system of international relations, which was developed in the wake of the Napoleonic wars, bolstered diplomacy among most European powers, it did not include the Ottoman Empire, thus leaving each nation to regulate its relations with the Ottomans separately.[281]

The Ottoman Empire was dubbed the "sick man" of Europe, and great powers around the world began developing contingency plans on how to best guide the falling empire to the fate that would best serve each nation's interest.[282] The Eastern Question was a sensitive subject throughout Europe and not often spoken of openly, while each nation sought to ensure that it would benefit most from the failing state.[283] Instead, nations, including Britain, Austria, and Russia, continued propping up the Ottoman Empire to secure favorable trade and security agreements.[284] As a result, Sultan Abdülmecid I entered into a series of private—and often conflicting—agreements with many European powers regarding rights and influences in the Ottoman Empire.

---

[276] In May 1850, Napoleon III announced authority over the Christian population of Palestine, which violated the London Straits Convention and induced the sultan of the Ottoman Empire to grant authority over Christian holy places in the Ottoman Empire, including those held by Russia (Schumacher, 2014, p. 69).

[277] Tsygankov, 2012; Edgar Anderson, "The Crimean War in the Baltic Area," *Journal of Baltic Studies*, Vol. 5, No. 4, 1974.

[278] Evans, 2023.

[279] Badem, 2010; Andrew Rath, *The Global Dimensions of Britain and France's Crimean War Naval Campaigns Against Russia, 1854–1856*, thesis, McGill University, 2012.

[280] Tsygankov, 2012.

[281] Tsygankov, 2012.

[282] Schumacher, 2014, p. 69.

[283] In fact, Britain and Russia both attempted to sow doubt in the mind of the sultan regarding their rival countries' intentions (G. H. Bolsover, "Lord Ponsonby and the Eastern Question (1833–1839)," *Slavonic and East European Review*, Vol. 13, No. 37, 1934).

[284] Tsygankov, 2012.

After Russia, many European stakeholders saw the UK as the second-strongest power in the region. London had an interest in gaining competitive advantage against Russia, maintaining geopolitical balance, and ensuring that the Ottoman Empire continued providing important trade opportunities for the British. Prior to the war, Russia and the Ottoman Empire traded significantly with the UK, but the Ottomans imported a larger number of British goods than Russian goods.[285] During this period, international affairs were becoming a larger part of the lives of average British citizens, and politicians were finding themselves beholden to the opinions of the masses, which were often swayed by increasingly anti-Russian newspaper campaigns.[286] Spurred by growing Russophobia as a result of Nicholas I's suppression of revolutions, these campaigns led British politicians to believe that Russia was gearing up to take Constantinople.[287] In terms of military might, the UK held a position of the "Two Power Standard" that sought to maintain a British navy that was equal in power to the second- and third-strongest navies combined, which were usually those of Russia and France.[288] The British army was significantly weaker and unable to hold its own against other major European powers.[289]

The French position on the Eastern Question leading up to the Crimean War is difficult to track because of domestic instability and internal turmoil within French politics; however, there was a growing desire to restore France's international prestige. France had gone through a period of successive political turmoil beginning with the fall of Napoleon in 1812, his brief return, the July Revolution of 1830, the Revolution of 1848 (which elected Napoleon III to the role of president), and the coup d'état that placed Napoleon III as emperor of France in 1852.[290] Napoleon III, who was in a precarious position domestically, was seeking an external diversion from internal conflict and wanted to restore the grandeur of France. Napoleon III also expressed concern that Nicholas I did not respect him as an emperor, instead addressing Napoleon III as a politician and, thus, did not recognize the legitimacy of his rule over France.[291] Because France had suffered a significant decline in international prestige, Napoleon III wished to embroil France in international affairs to assert French strength, build domestic support for his reign, and gain international respect.[292] Militarily, France had been fighting continuous colonial wars in Algeria since 1830, and its army was well equipped and had the experience to begin a new campaign.[293]

The positions that other nations took toward the Eastern Question were shaped primarily by a desire to maintain the geopolitical balance of the region and prevent any nation from carving up the Ottoman Empire in its favor alone. Austria had a series of private agreements with Russia and other European powers regarding the Eastern Question and the maintenance of the status quo. These

---

[285] Lambert, 2010.

[286] Rath, 2012.

[287] Bolsover, 1934; Tsygankov, 2012.

[288] Lambert, 2010.

[289] Lambert, 2010.

[290] Rath, 2012.

[291] Badem, 2010.

[292] Badem, 2010; Rath, 2012; Brison D. Gooch, "A Century of Historiography on the Origins of the Crimean War," *American Historical Review*, Vol. 62, No. 1, October 1956.

[293] James B. Agnew, "The Great War That Never Was: Crimea, 1853–1856," *Parameters*, Vol. 3, No. 1, 1973.

agreements were primarily aimed at avoiding conflict with France and maintaining connections with Russia for Austria's large Slavic population.[294] Prussia and other European nations had concluded that Russia was both capable of and intent on destroying the Ottoman Empire and occupying Constantinople.[295]

Takeover

On July 3, 1853, Russian forces entered the Danubian principalities of Moldavia and Wallachia to aid in quelling riots and began occupying the region.[296] Russia did not believe that any other European nations would become invested in the dispute. However, Russia misjudged Britain's tolerance for Russian occupation in the Balkans, Austria's willingness to partner with Russia, and the concerns about Russian naval presence in the Dardanelles Strait.[297]

France responded by partnering with Britain and moving a combined fleet through the Dardanelles.[298] Shortly after, the sultan declared war on Russia and sent troops across the Danube. However, Ottoman military capabilities could not stand up to those of Russia. In February 1854, France and Britain warned Russia to pull out of the Ottoman Empire.[299] After Russia refused in March 1854, Britain and France declared war on Russia. The war had many fronts, but one of the largest battles was the siege of Sevastopol on the Crimean Peninsula.[300] Although Russia had a standing army of 1.8 million regular troops, its long borders meant that only a fraction could be deployed to any one theater, with the largest contingency of 100,000 troops deployed to defend Sevastopol.[301] Following an 11-month siege, Sevastopol fell under heavy bombardment and because of a lack of supplies in September 1855. In December, Austria issued an ultimatum to cease fighting; because Tsar Nicholas I had died and was succeeded by his son, Alexander II, who was much less interested in propagating the war, a peace was reached in early 1856.[302]

Regional and Great Power Reactions

Despite a strong advantage against the Ottoman forces, the Russian military failed to hold out against the combined British and French forces because of a combination of low funds, internal corruption, a lack of international support, and a military spread too thin. Although the Russian army

---

[294] Badem, 2010.

[295] Gooch, 1956.

[296] Notably, the occupation did not specifically violate any active agreements and did not, on its own, mandate war (Evans, 2023; Gooch, 1956).

[297] Evans, 2023.

[298] Lambert, 2010.

[299] Scholars have argued that France and Britain entered the war only after it became clear that the Ottoman Empire would be unable to defend itself against Russia's advance (Mara Kozelsky, "The Crimean War, 1853–56," *Kritika: Explorations in Russian and Eurasian History*, Vol. 13, No. 4, 2012).

[300] Kozelsky, 2012.

[301] Tsygankov, 2012.

[302] Lambert, 2010.

was the largest army in Europe, it did not send a very large contingent to Crimea because Russia needed to maintain defenses along the border with Europe, especially along the Swedish, Prussian, and Austrian borders.[303] Additionally, Russia's navy fell behind the British and French fleet because of corruption and inefficiency in the naval structure.[304] This combination left Sevastopol relatively underdefended, leading it to become one of the most significant battlegrounds of the war. In the eyes of the Russian people, Russia's largest victory in the war was its ability to hold the defense of Sevastopol for 11 months. Even though the city eventually fell, Russia considered its defense one of the great moral victories of the era.[305] Alexander II did not place as much importance on the protection of Ottoman Christians as his father did and quickly sought peace with the coalition along the lines proposed by Austria, fearing that prolonging the war would lead to less favorable terms of peace.[306] Overall, although the Russians ultimately failed, domestically the Russian people perceived the war as a moral victory, and the successful defense of Sevastopol was considered more important to the Russian public than the defense of Ottoman Christians.

Throughout the war, the Ottoman forces suffered significant internal rivalry, desertion, poor equipment, and poor training. The Ottoman military structure of the time allowed for Muslims only to serve in the army, and higher ranks were granted based on favoritism, which led to rivalry and distrust between low-ranking soldiers and officers.[307] This feeling of rivalry was further exacerbated by Ottoman officers who, seeing the war as a means of personal enrichment, took supplies from their troops, leaving their soldiers without proper equipment, food, and clothing. Ottoman troops also did not get along with the inhabitants of many areas that they were meant to defend, and local protests against the Ottoman requisitions of supplies led to the violent suppression of the locals.[308] In turn, these problems led to a high rate of desertion, which allowed officers to pocket the funds allocated for the missing soldiers.[309] Additionally, the Ottomans were not well liked by the British and French, who struggled to communicate with Ottoman officers because of the language barrier. British and French troops treated Ottoman soldiers poorly, often assigning them the lowest-ranking jobs.[310] These issues meant that the Ottomans had to rely on British and French troops for most operations, which led to further costs for the British and French.

The British participation in the Crimean War was driven primarily by a desire to maintain geopolitical balance and retain economic agreements with the Ottomans. British participation was also colored by domestic opinion because of the rise of media attention over the course of the war. The British military was prepared for a primarily naval battle and had deployed only 40,000 army troops (compared with approximately 240,000 ground troops for France).[311] The British strategy relied on

---

[303] Badem, 2010.

[304] Rath, 2012.

[305] Pierre Malinowski and Elena V. Linkova, "The Crimean War of 1853–1856 in the Memory Space of Russia and France," *RUDN Journal of Russian History*, Vol. 20, No. 2, 2021; Tsygankov, 2012.

[306] Lambert, 2010.

[307] Badem, 2010.

[308] Orlando Figes, *The Crimean War: A History*, Metropolitan Books, 2010.

[309] Badem, 2010.

[310] Figes, 2010.

[311] Lambert, 2010.

blockades, the cutting of logistic supply lines, and naval attacks throughout 1854; later, the British focused on Sevastopol and protecting trade routes to British India.[312] Despite military blockades, the UK kept trade routes with Russia open for commercial products, as well as several with both civilian and military application, including hemp and tallow.[313]

Domestic British opinion played an extremely pivotal role in the actions of the UK, spearheading British participation and instigating resignations from both military leaders and the prime minister. Public opinion was shaped largely by the British media attention: Although the public supported the war effort, key elements of the British government did not wish to engage in, and subsequently was blamed for, participation in the war. Specifically, the British prime minister did not support the war effort, but Parliament did, citing public opinion and media.[314] During the war, Britain's domestic audience found British naval and land campaigns to be severely lacking based on the volume, and often bias, in coverage of the war by British media.[315] Because of this poor press, Parliament began an inquiry into the conduct of the war, and the prime minister and first lord of the admiralty were both forced to resign.[316] Their replacements failed to improve in performance.

France's motivation for entering the war was to obtain notable victories to shore up domestic favor for Napoleon III and build international prestige. The most significant victory for France was the siege of Sevastopol; when the city fell, France considered the war to be over and expressed disinterest in achieving further strategic objectives proposed by the British.[317] For France, the battle of Sevastopol represented its regained prestige and was cited as a symbol of the perseverance and heroism of French soldiers.[318] Ultimately, France achieved its objective for prestige without significant domestic backlash or international consequences, although France's cooperation with Britain did not last long.

France and the UK disagreed on most issues, and the alliance between them was fraught with miscommunication and distrust; this bickering led to poor military decisions that were attempts to appease both parties. Prior to the outbreak of war, Britain and France had been rivals in most European conflicts for the preceding five centuries.[319] However, Britain had grown increasingly concerned about Russian ambitions in the Balkans and was painfully aware that the British army would fare poorly against Russian troops. The UK sought a coalition—rather than an alliance—with France because a coalition was seen as more temporary and focused on a single objective.[320] However, although the agreement was not finalized until April 10, 1854, an alliance was officially formed two

[312] Lambert, 2010.

[313] Jan Martin Lemnitzer, "'More Serious than the Eastern Question Itself'—the Crimean War Compromise," *Power, Law and the End of Privateering*, Palgrave Macmillan, 2014.

[314] Rath, 2012.

[315] This includes the publication of Alfred Tennyson's poem, "The Charge of the Light Brigade," which mourned a fatal cavalry charge and condemned it as a British strategic failure, despite the fact that—while the charge created significant casualties—the operation successfully scattered Russian troops (Figes, 2010).

[316] Figes, 2010; Lambert, 2010.

[317] Lambert, 2010.

[318] Malinowski and Linkova, 2021.

[319] Lambert, 2010.

[320] Lambert, 2010.

weeks after both Britain and France had declared war on the Russian Empire.[321] Throughout the war, there were competing opinions on objectives and battle strategies, with British and French military leaders unwilling to take orders from their counterparts and distrustful of each other's motives, leading to half-measures and delays that impeded the effectiveness of alliance troops.[322] Despite these issues, scholars noted that Ottoman military failures were the primary cause of difficulties in the field and could not be attributed to the coalition.[323] The British-French alliance was fraught with poor coordination, clashing objectives, and distrust; however, their complementary capabilities, combined with the fact that Russia's defense was so strapped for cash, allowed the alliance to succeed. Although the alliance crumbled following the war, it did prevent Russia from gaining a further foothold in Ottoman territory.

During the conflict, several other nations attempted to use the distraction of the conflict in Europe to pursue their own interests in the region. In the Balkans, supporters of the Slavic identity across the region labeled the Crimean War as holy and treated Russia's interest in protecting the Ottoman Christians as a necessary step to supporting the ethnic divide in the region, calling on Russia to employ more-aggressive tactics or even invade Austria.[324] Initially, Iran sought to take advantage of the occupation of Ottoman forces elsewhere to take Baghdad, moving its troops to the Ottoman border. However, Britain sent delegations to prevent Iran's invasion.[325] Sweden and Denmark became concerned about the safety of commercial vessels and the flow of trade, insisting on agreed-upon rules with the alliance and Russia for "prize vessels" and ports and to ensure that trade continued.[326] Fighting also spilled over to areas of the Pacific where other neutral powers sought to avoid shipping disruptions.[327] Finally, the small kingdom of Sardinia-Piedmont joined the war shortly before the war's conclusion in 1855 with a contingent of 10,000 troops in an effort to gain concessions through their participation.[328] These incidents show that when the major regional powers were distracted in a larger conflict, not only did clashes spill out to other regions but some neutral powers also sought a return to the status quo. At the same time, others attempted to enrich their prospects either by joining the side they predicted would win or by taking advantage of the chaos to achieve other objectives.

Aftermath

After the fall of Sevastopol, Austria was brought in as a third party to mitigate British and French distrust and broker a lasting agreement. Austria issued an ultimatum that it would enter the fight on

[321] Rath, 2012.

[322] Notably, Napoleon III sought to personally lead his troops and often sent last-minute or conflicting orders, which led the French troops to act out of sync with their predetermined plans with their British counterparts (Figes, 2010).

[323] Badem, 2010.

[324] Tsygankov, 2012; Figes, 2010.

[325] Badem, 2010.

[326] Lemnitzer, 2014.

[327] Rath, 2012.

[328] Agnew, 1973.

the side of the alliance if Russia did not agree to a peace.[329] This ultimatum, combined with British threats of building 300 more gunboats to support the 32 vessels already engaged in the war, led Russia to agree to negotiate peace.[330] The war ended with two main agreements: the 1856 Peace of Paris and the 1856 Paris Declaration Respecting Maritime Law.[331] The first agreement centered on maintaining geopolitical balance and increasing British and French access to new economic markets in the east. As part of the agreement, Russia relinquished protection rights to Turkish Orthodox Christians, neutralized the Black Sea, surrendered southern Bessarabia, and opened the Danube River to international trade.[332] Russia's concessions following the war were relatively tame, and Tsar Alexander II came to the proceedings as an equal participant rather than as a vanquished combatant.[333] The second agreement established international maritime laws for conflict and peace to shore up confidence in trade by sea. The laws abolished privateering and established rules regarding neutral flags, captured cargo, and trade during conflict.[334] The agreement was signed by most European states, although the United States and Spain refrained from signing.[335]

Regional and Great Power Alignment

Following the conclusion of the conflict, Russia was no longer viewed as a strong international partner, and France had a short-lived rise in international prestige. Although Britain did not initially appear to gain significantly from the war, it came to be seen as the only member of the Concert of Europe with global reach that had sought to maintain the status quo and preserve order through the creation of the new international laws.[336] In contrast, Russia suffered because its military was shown to be weak, it lost the right to protect Ottoman Christians, it struggled with rising economic issues as a result of war debt, and it was unable to maintain peace, engaging in several smaller clashes in the Balkans in the subsequent decades.[337] Finally, although France initially gained prestige through its victories, these were quickly overshadowed by conflicts with Austria in 1859 and Mexico in 1862.[338]

Shortly after the Crimean War, many of the smaller kingdoms of both Germany and Italy were unified under the king of Prussia and House of Savoy. Many scholars attribute this to fear of further war and the sentiment that the European powers had not fully settled their disputes in the Crimean War.[339] In fact, scholars argued that the threat of war rose—rather than diminished—following the

---

[329] Badem, 2010.

[330] Rath, 2012.

[331] Treaty of Paris, signed at Paris, March 30, 1856; Declaration Respecting Maritime Law, signed at Paris, April 16, 1856.

[332] Harold Temperley, "The Treaty of Paris of 1856 and Its Execution," *Journal of Modern History*, Vol. 4, No. 3, December 1932.

[333] Kozelsky, 2012; Rath, 2012.

[334] Declaration Respecting Maritime Law, signed at Paris, France, April 16, 1856.

[335] Rath, 2012.

[336] Schumacher, 2014.

[337] Rath, 2012; Evans, 2023.

[338] Evans, 2023.

[339] Agnew, 1973; Schumacher, 2014; Tsygankov, 2012.

Crimean War and that the Eastern Question was not solved but rather delayed, with the potential to cause a full European war.[340]

Key Factors

The key factors that affected the course of events were as follows:

- *Alliance of mutual concern:* Despite the historical competition between Britain and France, their mutual concern about Russia gaining territory from the flagging Ottoman Empire led to a reluctant cooperation. However, the alliance was plagued with distrust and conflicting goals.
- *Declining empire:* The Ottoman Empire was in a state of protracted decline because of a myriad of factors, including a failing economy, growing inequality, and internal corruption. These factors contributed to an ineffective and outdated military that required outside assistance to defend against an invasion.
- *Poor communication:* The Russian tsar sought primarily to put down rebellions and preserve the monarchical structures of Europe. However, Nicholas I failed to persuade concerned regional nations that Russia did not seek territorial gains or monetary advantages.

Conclusion

Although the Crimean War was initially started because of a small debate regarding jurisdiction over religious groups, it quickly swelled into a Europe-wide conflict as the decline of the Ottoman Empire brought into question the distribution of former Ottoman holdings. Throughout the conflict, regional European powers looked on with trepidation at the tenuous political balance and sought assurances that trade would continue unimpeded and the conflict would not spill over into adjacent regions. Britain and France managed to hold a contentious and fraught alliance through the course of the war, but the alliance did not last beyond the conflict because deep-seated distrust and suspicion plagued the relationship. In fact, the long-lasting aspect of the Crimean War conflict was the establishment of international maritime privateering laws.[341] The principles of the Paris Declaration have been reaffirmed and expanded in subsequent treaties and are a basis of existing international maritime law. The events of the Crimean War were part of a bigger issue regarding the future of Europe in the wake of the failing Ottoman Empire, which finally fell apart after World War I.

This case is instructive for contemporary territorial-revision scenarios because it shows the effectiveness of a coalition of nations acting based on mutual concerns about an invader. In the case of the Crimean War, the Russian military was independently stronger than the British army and the French navy. By operating together, France and Britain were able to overcome their weaknesses and exceed the strength of the Russian forces. This indicates the importance of coalitions in countering a large military. Additionally, the alliance was formed partially because of concerns about the balance of power and a belief that Russia would expand further into Europe and the Middle East if unchecked in

[340] Agnew argued that, had Russia been able to court an equal alliance and its technology had been more advanced, the "powder keg" of European distrust would have led to a much larger conflict (Temperley, 1932; Agnew, 1973; Rath, 2012).

[341] Declaration Respecting Maritime Law, 1856.

Crimea, which indicates that regional stakeholders can be spurred to action and cooperation by concerns about an invader will engage in further acts of aggression to expand its influence. Regional countries may be likelier to defend a weaker nation if they perceive that an aggressive nation will seek further expansion.

Finally, maritime law played a large role in the negotiations following the Crimean War. This is still an issue of ongoing importance to the global economy, with potential disruptions to shipping in the Strait of Hormuz, the Bab-el-Mandeb, and the Strait of Malacca and through the South China Sea. Free access to shipping and trade routes were critical to the decisionmaking calculus among Western European states, and the possibility that Russia could close these routes helped spur those countries to counter Russian expansion. Such considerations are still likely to continue to play an important role given the globalization of the economy and the importance of international trade and shipping.

Chapter 7

Cross-Case Comparisons and Takeaways

There are a series of throughlines that emerge from the historical case studies discussed in this report. First, fears of successive invasions drove regional behavior in all cases. After the Soviet invasion of Czechoslovakia in 1968, some communist states moved to balance Soviet strength. They were assisted in this effort by China's split from the USSR because of their own fear of being subjugated by Moscow. In addition, this split drove great powers outside the region to put their thumbs on the scale in favor of balance. Iraq's takeover of Kuwait meant that Iraq could more easily seize the main Saudi oil fields, potentially concentrating a significant share of the global oil supply in Saddam's possession. Similarly, the Russian seizure of Ottoman territory extended St. Petersburg's influence in eastern Europe, threatening the regional balance of power.

Revisionist powers sought to allay these fears in a variety of ways. The Soviet Union dampened Western concerns about spillover by emphasizing that the Brezhnev Doctrine applied only to already existing communist countries. The United States and others treated the issue as an internal matter for the Eastern bloc. That said, Soviet conduct inspired a reaction within that bloc, with Mao in particular leveraging their actions for his own domestic political purposes: to vie with Russia over leadership of the international communist movement and to prevent Moscow from launching a similar intervention against Beijing.[342] The Japanese also explicitly avoided attacking cities with large numbers of foreigners in its invasion of Manchuria in an effort to prevent more-concerted responses by great powers. Beijing sought to alleviate international concerns about its incorporation of Hong Kong by emphasizing its status as an SAR, although subsequent actions by the PRC significantly eroded trust in their intentions.

Second, the continuing and credible involvement of great powers was a significant factor in every case. This includes both the presence and absence of great powers. For example, after the Japanese invasion of Manchuria in 1931, the League of Nations repeatedly called on the British and Americans to intervene, but no intervention was forthcoming. As a result, the Japanese march forward continued until Japan's attack on U.S. territory in 1941. Contrarily, the U.S. decision to intervene after the Iraqi invasion of Kuwait in 1990 showed how a decisive great power intervention can determine the course of a conflict and the alignment patterns of cooperation in its aftermath. Great powers often have to assure third parties affected or threatened by the original territorial revision that their security interests are still recognized and protected. Saudi Arabia is the clearest example from among the cases, but Austria's participation in the Crimean War was also prompted in part by a similar need for

[342] Chen Jian, *Mao's China and the Cold War*, University of North Carolina Press, 2001, Chapter 3.

assurance.[343] Conversely, the Soviet invasion of Czechoslovakia illustrates how not providing assurances (at least in the short term) can be strategic. Allowing China to confront the Soviet Union with only some Eastern bloc support exacerbated the Sino-Soviet split and increased China's perceptions of threat and sense of isolation. In turn, the Sino-Soviet split facilitated the Sino-U.S. rapprochement, dividing and substantially weakening the communist bloc.

Third, regional and international organizations can play a key role in the conflict and the alignment of states afterward. In the UK's cession of Hong Kong in 1997, the UN gave its imprimatur to the handover because the UN saw it as a continuation of the decolonization movements that began after World War II.[344] Similarly, the Soviet invasion of Czechoslovakia was also undertaken under the umbrella of the Warsaw Pact, which provided the operation with additional legitimacy. In contrast, the Iraqi invasion of Kuwait was quickly condemned by the UN, and both Egypt and Saudi Arabia heavily lobbied the Arab League to condemn it as well. After the Arab League's condemnation, Iraq was isolated both regionally and globally. Iraq had few allies which it could rely on for support within these organizations. The post–World War II norm for territorial integrity has led to an unprecedentedly high survival rate for small states.[345]

However, weak institutions can prevent timely and effective responses to revisionism. France and the UK struggled to maintain their coalition against Russia during the Crimean War. They failed to resolve their underlying sources of tension, and the coalition's subsequent collapse meant that that alignment had to be rebuilt the next time the Eastern Question reemerged. Similarly, communist international organizations—dominated as they were by the Soviet Union—provided no support to Czechoslovakia. Additionally, the Manchuria case illustrates how failing international institutions can facilitate further territorial aggression. The Western great powers repeatedly placed Japan in a subordinate status in their international agreements. Moreover, their failure to uphold key institutions—such as the treaty port system—meant that Tokyo gained little benefit from following global norms, rules, and procedures. The U.S. rejection of League of Nations membership also hamstrung the league's efforts to mediate the Sino-Japanese conflict because it simply was not in the room in which those discussions were occurring. This culminated in Japan quitting the league and renewing its territorial aggression four years later in the second Sino-Japanese War, which expanded into World War II.

Fourth, economic links played a critical role in every conflict. For instance, a key impetus for British and French intervention in the Crimean War was fear of being economically shut out of the region by an ascendant Russia. The peace that came after the conflict made this motivation clear because it guaranteed economic access and forbade acts of piracy that Russia could use to imperil access to that region in the future. The Japanese invasion of Manchuria was also highly related to economics, with the Japanese needing access to more natural resources than existed in its home islands. The lack of economic links with the Chinese interior, along with their own economic struggles, was also a key reason that the British and Americans refused to intervene. Fear of Iraqi

[343] Stephen Van Evera, "American Intervention in the Third World: Less Would Be Better," *Security Studies*, Vol. 1, No. 1, Autumn 1991.

[344] Cockram, 2018.

[345] Tom Long, *A Small State's Guide to Influence in World Politics*, Oxford University Press, 2022.

control over a large portion of the world's oil supply was also fundamental to global decisions to denounce the invasion and support its rollback.

Last, the treatment of populations ruled after a territorial takeover played a key role in every situation. How the PRC has treated people in Hong Kong since the handover continues to be an issue to this day and stands as an example to both regional and global audiences about the authoritarian nature of the Beijing government. The harshness of the Soviet intervention in Czechoslovakia also convinced some communist states to distance themselves from Moscow. Stories related to the treatment of Kuwaitis by Iraqi forces galvanized Arab and global public opinion against the Iraqi regime after the invasion, which was then flipped on its head when the United States was accused of human rights abuses owing to the sanctions on Iraq following the war. Even in the Crimean War, the treatment of Christian populations in Ottoman territory played a key role in the motivation of outside powers to intervene.

The analysis of these five historical case studies across different periods, different regions, and with varying levels of fighting shows that they have five key factors in common, which are shown in Table 7.1.

Table 7.1. Key Factors in Determining Regional and International Alignment in Historical Cases

Historical Case	Fears of Successive Takeovers	Continued, Credible Involvement from Great Power (or Lack Thereof)	International and/or Regional Organization Legitimacy	Economic Links	Harsh Treatment of Target Population
UK cession of Hong Kong 1997	✓	✓	✓	✓	✓
USSR invasion of Czechoslovakia 1968	✓	✓		✓	✓
Japan invasion of Manchuria 1931	✓	✓	✓	✓	✓
Iraq invasion of Kuwait 1990	✓	✓	✓	✓	✓
Crimean War 1853–1856	✓	✓		✓	✓

We can draw a few key conclusions from these cases. These findings can inform U.S. policy in the event of territorial aggression, for example, in the case of the Russian war on Ukraine that began in 2022, regional tensions around the South China Sea, or potential invasion of Taiwan. Revisionism—in particular, the threat of spillover—often prompts not only the victim but also third parties to focus on their defensive needs. Having witnessed international institutions, great powers, and other countries fail to prevent a territorial seizure, these states scramble to shore up their own security by purchasing arms, tightening security cooperation, and seeking firmer assurances that their security interests will be protected. In the absence of such assurances, regional states may decide that their interests are best protected by siding with the revisionist to avoid punishment. As the primary security partner for many countries, the United States can shape that response to bolster combined intelligence gathering and sharing, military planning, and warfighting capabilities. In particular, the Department of the Air Force (DAF) can work to ensure that regional military modernization and arms acquisition efforts align with its plans and strategies, as well as to secure for the U.S. Air Force the necessary basing, access, and overflight rights in the event of and run-up to a conflict.

This requires a deeper and more nuanced understanding of states' priorities. How does spillover threaten their specific interests? To what extent do other countries share those concerns, and, therefore, how broad is the coalition that can respond? Critically, how do threatened states evaluate the credibility and capabilities of great powers in the wake of territorial revision? For the DAF, is this purely a question of forward and prepositioned forces, or can political commitments and operational coordination substitute for some of these measures?

International institutions and norms play key roles in answering these questions. Norms act as focal points for evaluating a conqueror's justifications for territorial seizure. States observe the extent to which a conqueror adheres—and, especially, fails to adhere—to those norms as signals of further territorial ambitions. Efforts at norm enforcement and framing can drive third-party threat perceptions and provide the United States with further opportunities to guide regional security planning. Institutions can facilitate communication and coalition-building, reducing transaction costs and speeding a collective response. In particular, DAF engagement with alliance partners provides a pathway to affect this normative and institutional coordination. For instance, such programs as the International Military Education and Training, hosting liaison officers, and other kinds of exchange programs develop shared norms that are integral to this process, not just high-level diplomatic bodies, such as the UN. Therefore, the DAF can play a key role in setting the stage for common norms and collaboration in international bodies.[346]

---

[346] Carla Martinez Machain, "Exporting Influence: US Military Training as Soft Power," *Journal of Conflict Resolution*, Vol. 65, Nos. 2–3, 2021.

Appendix

Cases of Territorial Conquest (1946–2017)

Table A.1. Cases of Territorial Conquest

Perpetrator	Victim	Territory	Year	Victim Held Territory (Immediately After)	Victim Held Territory (Ten Years Later)	Third Party–Defended Victim	Victim in Defense Pact
North Korea	South Korea	South Korea	1950	Yes	Yes	Yes	No
Bulgaria	Greece	Gamma Island	1952	Yes	Yes	No	Yes
Saudi Arabia	UK	Al-Buraimi	1952	Yes	Yes	No	Yes
Thailand	Cambodia	Preah Vihear	1954	No	No	No	No
South Korea	Japan	Dokdo (Takeshima) Islands	1954	No	No	No	Yes
India	Portugal	Dadra and Nagar Haveli	1954	No	No	No	Yes
Israel	Egypt	Sinai; Gaza	1956	Yes	Yes	No	Yes
Nicaragua	Honduras	Mocorón	1957	Yes	Yes	No	Yes
Morocco	Spain	Ifni (p)	1957	Yes	No	No	No
Egypt	Sudan	Halaib Triangle	1958	Yes	Yes	No	No
India	China	Longju, Kongka Pass	1959	Yes	Yes	No	Yes
India	Portugal	Goa	1961	No	No	No	Yes
India	China	Ladakh; Arunachal Pradesh borders (p)	1962	Yes	Yes	No	Yes
Morocco	Algeria	Colomb-Béchar; Tindouf	1963	Yes	Yes	Yes	No
Pakistan	India	Rann of Kutch (p)	1965	No	No	No	No
Pakistan	India	Akhnoor	1965	Yes	Yes	No	No

Perpetrator	Victim	Territory	Year	Victim Held Territory (Immediately After)	Victim Held Territory (Ten Years Later)	Third Party–Defended Victim	Victim in Defense Pact
Venezuela	Guyana	Ankoko Island	1966	No	No	No	No
Israel	Egypt	Sinai; Gaza	1967	No	No	Yes	Yes
China	Russia	Damansky (Zhenbao) Island	1969	No	No	No	Yes
El Salvador	Honduras	Gulf of Fonseca islands; six border pockets	1969	Yes	Yes	No	Yes
Argentina	Uruguay	Timoteo Domínguez (Punta Bauza)	1969	No	Yes	No	Yes
Yemen People's Republic	Saudi Arabia	Al-Wadiah	1969	Yes	Yes	No	Yes
Iran	UK	Abu Musa; Greater and Lesser Tunbs	1971	No	No	No	Yes
Philippines	China	Spratly Islands (p)	1971	No	No	No	Yes
Yemen	Yemen People's Republic	Kamaran	1972	No	No	No	Yes
Iraq	Kuwait	Al-Samita	1973	Yes	Yes	No	Yes
Egypt	Israel	Sinai (p)	1973	Yes	No	No	No
Syria	Israel	Golan Heights (p)	1973	Yes	Yes	No	No
Turkey	Cyprus	Northern Cyprus	1974	No	No	Yes	Yes
China	Republic of Vietnam	Paracel Islands (p)	1974	No	No	No	No
Vietnam	Republic of Vietnam	South Vietnam	1974	No	No	No	No
Morocco	Spain	Western Sahara	1975	No	No	No	No
Cambodia	Vietnam	Phu Quoc; Tho Chu	1975	Yes	Yes	No	No
Indonesia	Timor-Leste	Timor-Leste	1975	No	No		No
Argentina	UK	Thule Island	1976	No	Yes	No	Yes
Somalia	Ethiopia	Ogaden	1977	Yes	Yes	Yes	Yes

Perpetrator	Victim	Territory	Year	Victim Held Territory (Immediately After)	Victim Held Territory (Ten Years Later)	Third Party–Defended Victim	Victim in Defense Pact
Cambodia	Vietnam	Tay Ninh, Ha Tien, adjacent areas	1977	Yes	Yes	No	No
Uganda	Tanzania	Kagera Salient	1978	Yes	Yes	Yes	No
Iraq	Iran	Khuzestan	1980	Yes	Yes	No	No
Ecuador	Peru	Cordillera del Cóndor (p)	1981	Yes	Yes	No	Yes
Argentina	UK	Falkland (Malvinas) Islands	1982	Yes	Yes	No	Yes
Nigeria	Chad	Islands in Lake Chad	1983	No	No	No	No
Malaysia	China	Spratly Islands (p)	1983	No	No	No	Yes
Thailand	Laos	Sawang; Klang; Mai	1984	Yes	Yes	No	No
India	Pakistan	Siachen Glacier	1984	No	No	No	Yes
India	China	Thag La Ridge	1984	Yes	Yes	No	Yes
Mali	Burkina Faso	Agacher Strip	1985	Yes	No	No	Yes
China	India	Thag La Ridge	1986	No	Yes	No	No
Qatar	Bahrain	Fasht al Dibal	1986	Yes	Yes	No	Yes
Thailand	Laos	Rom Klao	1987	Yes	Yes	No	No
Nigeria	Cameroon	Islands in Lake Chad	1987	No	Yes	No	No
China	Vietnam	Spratly Islands (p)	1988	No	No	No	No
Peru	Ecuador	Cordillera del Cóndor	1990	Yes	Yes	No	Yes
Iraq	Kuwait	Kuwait	1990	Yes	Yes		Yes
Botswana	Namibia	Kasikili/Sedudu Island	1991	No	No	No	No
Nigeria	Cameroon	Diamond and Jabane Islands; Bakassi Peninsula	1993	No	No	No	No

Perpetrator	Victim	Territory	Year	Victim Held Territory (Immediately After)	Victim Held Territory (Ten Years Later)	Third Party–Defended Victim	Victim in Defense Pact
Egypt	Sudan	Halaib Triangle	1995	No	No	No	Yes
Ecuador	Peru	Cenepa River (p)	1995	Yes	Yes	No	Yes
China	Philippines	Spratly Islands (p)	1995	No	No	No	Yes
Eritrea	Yemen	Hanish Islands	1995	No	Yes	No	Yes
Vietnam	Cambodia	Small areas in Svay Rieng, Kampong Cham, Prey Veng Provinces	1995	No	No	No	No
Greece	Turkey	Imia (Kardak)	1996	Yes	Yes	No	Yes
Eritrea	Ethiopia	Badme	1998	Yes	Yes	No	Yes
Pakistan	India	Kargil	1999	Yes	Yes	No	No
Morocco	Spain	Parsley (Perejil) Island	2002	Yes	Yes	No	Yes
Eritrea	Djibouti	Ras Doumeira (p)	2008	No	No	No	Yes
Nicaragua	Costa Rica	Calero Island (p)	2010	No		No	Yes
Russia	Ukraine	Crimea	2014	No		No	No
Eritrea	Djibouti	Doumeira (p)	2017	No		No	Yes
India	China	Doklam	2017	Yes		No	Yes

SOURCE: Altman, 2020.

NOTE: (p) indicates partial conflict. The data run from 1918 to 2017: This is the post–World War II set of conquest attempts requested by the sponsor. Because the dataset ends in 2017, conquest attempts in the prior ten years do not have codes for whether the victim held the territory ten years later, except when the outcome had already been determined (an example is Eritrea's attempt on Djibouti in 2008). Other missing data can be seen in the "Third Party–Defended Victim" column. These are often the result of indeterminant coding decisions, such as where third parties intervened but only *after* a successful conquest, such as Kuwait in 1990.

Abbreviations

ACC	Arab Cooperation Council
CCP	Chinese Communist Party
DAF	Department of the Air Force
GCC	Gulf Cooperation Council
NATO	North Atlantic Treaty Organization
PLO	Palestine Liberation Organization
PRC	People's Republic of China
SAR	special administrative region
UK	United Kingdom
UN	United Nations
UNSC	United Nations Security Council
USSR	Union of Soviet Socialist Republics

References

Agnew, James B., "The Great War That Almost Was: Crimea, 1853–1856," *Parameters*, Vol. 3, No. 1, 1973.

Alston, Philip, "The Security Council and Human Rights: Lessons to Be Learned from the Iraq-Kuwait Crisis and Its Aftermath," *Australian Year Book of International Law*, Vol. 13, 1992.

Altman, Dan, "Replication Data for: 'The Evolution of Territorial Conquest After 1945 and the Limits of the Norm of Territorial Integrity,'" Harvard Dataverse, Vol. 1, 2020. As of March 10, 2025: https://dataverse.harvard.edu/dataset.xhtml?persistentId=doi:10.7910/DVN/XLYTC7

Altman, Dan, "Codebook for the Modern Conquest Dataset Version 2.2," January 10, 2021.

Amnesty International, "Iraq/Occupied Kuwait: Human Rights Violations Since 2 August," December 19, 1990.

Anderson, Edgar, "The Crimean War in the Baltic Area," *Journal of Baltic Studies*, Vol. 5, No. 4, 1974.

"August 2, 1990: Iraq Invades Kuwait," webpage, BBC, undated. As of March 10, 2025: http://news.bbc.co.uk/onthisday/hi/dates/stories/august/2/newsid_2526000/2526937.stm

Badem, Candan, *The Ottoman Crimean War (1853–1856)*, Brill, 2010.

Barnes, Harry, interview with Charles Stuart Kennedy, Association for Diplomatic Studies and Training Foreign Affairs Oral History Project, April 25, 2001.

Beasley, W. G., *Japanese Imperialism: 1894–1945*, Oxford University Press, 1987.

Benčik, Antonín, et al., "Report by a Czechoslovak State Security (StB) Official on the Arrest of Dubček and Other Members of the CPCz CC Presidium, August 21, 1968," in Jaromír Navrátil, Antonín Benčík, Václav Kural, Marie Michálková, and Jitka Vondrová, eds., *The Prague Spring, 1968: A National Security Archive Documents Reader*, Central European University Press, [August 21, 1968] 2006.

Bolsover, G. H., "Lord Ponsonby and the Eastern Question (1833–1839)," *Slavonic and East European Review*, Vol. 13, No. 37, July 1934.

Bowie, Wendy Dullea, "The Effect of the Tiananmen Square Massacre upon Negotiations for the Draft Basic Law of the Hong Kong Special Administrative Region Comment," *Penn State International Law Review*, Vol. 8, No. 2, 1990.

Brezhnev, Leonid, "Transcript of Leonid Brezhnev's Telephone Conversation with Alexander Dubcek, August 13, 1968," in Jaromír Navrátil, Antonín Benčík, Václav Kural, Marie Michálková, and Jitka Vondrová, eds., *The Prague Spring, 1968: A National Security Archive Documents Reader*, Central European University Press, [August 13, 1968] 2006.

Burke, Paul J., "Economic Growth and Political Survival," *B. E. Journal of Macroeconomics*, Vol. 12, No.1, 2012.

Burkman, Thomas W., *Japan and the League of Nations: Empire and World Order, 1914–1938*, University of Hawaii Press, 2007.

Carapico, Sheila, "Yemen: Unification and the Gulf War," *Middle East Research and Information Project*, Vol. 170, May–June 1991.

Chan, Ming K., "Hong Kong: Colonial Legacy, Transformation, and Challenge," *Annals of the American Academy of Political and Social Science*, Vol. 547, No. 1, September 1996.

Chan, Ming K., "The Legacy of the British Administration of Hong Kong: A View from Hong Kong," *China Quarterly*, Vol. 151, September 1997.

Chapman, Michael E., "Fidgeting over Foreign Policy: Henry L. Stimson and the Shenyang Incident, 1931," *Diplomatic History*, Vol. 37, No. 4, September 2013.

Chen Jian, *Mao's China and the Cold War*, University of North Carolina Press, 2001.

Cheng, Joseph S., "The Future of Hong Kong: Surveys of the Hong Kong People's Attitudes," *China Journal*, Vol. 12, July 1984.

"Chinese Reactions to the Invasion of Czechoslovakia," *Studies in Comparative Communism*, Vol. 2, No. 1, 1969.

Cockram, Rebekah C., *Cession and Retreat: Negotiating Hong Kong's Future, 1979–1984*, thesis, University of North Carolina at Chapel Hill, 2018.

"Confrontation in the Gulf; Excerpts from Iraqi Document on Meeting with U.S. Envoy," *New York Times*, September 23, 1990.

Convention Between the United Kingdom and China, Respecting an Extension of Hong Kong Territory, signed at Peking, June 9, 1898.

Cuffe, Honae, "The Manchurian Crisis and the Genesis of Australian Foreign Policy," *The Strategist*, Australian Strategic Policy Institute, January 21, 2022.

Current, Richard N., "The Stimson Doctrine and the Hoover Doctrine," *American Historical Review*, Vol. 59, No. 3, April 1954.

Dagati, Patricia A., "Hong Kong's Lost Right to Self-Determination: A Denial of Due Process in the United Nations," *NYLS Journal of International and Comparative Law*, Vol. 13, No. 1, 1992.

Declaration Respecting Maritime Law, signed at Paris, France, April 16, 1856.

Deletant, Dennis, "New Evidence on Romania and the Warsaw Pact, 1955–1989," Wilson Center, July 7, 2011. As of March 12, 2025: https://www.wilsoncenter.org/publication/new-evidence-romania-and-the-warsaw-pact-1955-1989

Donati, Jessica, "U.S. Officially Declares That Hong Kong Is No Longer Autonomous," *Wall Street Journal*, May 27, 2020.

Dris-Aït-Hamadouche, Louisa, and Yahia H. Zoubir, "The US-Saudi Relationship and the Iraq War: The Dialectics of a Dependent Alliance," *Journal of Third World Studies*, Vol. 24, No. 1, Spring 2007.

Elliott, Mark C., "The Limits of Tartary: Manchuria in Imperial and National Geographies," *Journal of Asian Studies*, Vol. 59, No. 3, August 2000.

Evans, Alexandra T., *Alternative Futures Following a Great Power War*: Volume 2, *Supporting Material on Historical Great Power Wars*, RAND Corporation, RR-A591-2, 2023. As of February 12, 2024: https://www.rand.org/pubs/research_reports/RRA591-2.html

Evans, Rowland, and Robert Novak, "An Erosion of Support," *New York Times*, September 20, 1990.

Executive Order 13936, "The President's Executive Order on Hong Kong Normalization," Executive Office of the President, July 14, 2020.

Fatton, Lionel P., "The Manchurian Crisis as an Exogenous Shock, 1931–1933," in *Japan's Rush to the Pacific War: The Institutional Roots of Overbalancing*, Palgrave Macmillan, 2023.

Figes, Orlando, *The Crimean War: A History*, Metropolitan Books, 2010.

Fong, Brian C. H., "Great Power Competition in Non-Sovereign Entities: US–China Tug-of-War over Hong Kong, 1950–2020," *Pacific Focus*, Vol. 37, No. 2, August 2022.

"From Spring, into a Long Winter's Night: The Czechoslovakian Crisis of 1968," Cryptologic Almanac 50th Anniversary Series, DOCID: 3519758, last reviewed February 28, 2003.

Frommer, Fred, "When Soviet-Led Forces Crushed the 1968 'Prague Spring,'" History, March 14, 2022. As of February 12, 2024:
https://www.history.com/news/prague-spring-czechoslovakia-soviet-union

Garver, John W., "China and American Hegemony in the Persian Gulf," in *China's Quest: The History of the Foreign Relations of the People's Republic of China*, Oxford University Press, 2018.

Goldfrank, David M., *The Origins of the Crimean War*, Routledge, 1994.

Gooch, Brison D., "A Century of Historiography on the Origins of the Crimean War," *American Historical Review*, Vol. 62, No. 1, October 1956.

Goshko, John M., and R. Jeffrey Smith, "State Dept. Assailed on Iraq Policy," *Washington Post*, September 19, 1990.

Government of Hong Kong, "Summary of Expenditure Estimates," 2021.

Government of Hong Kong, "Summary of Expenditure Estimates," 2023.

Government of Hong Kong, "Summary of Expenditure Estimates," 2024a.

Government of Hong Kong, "Safeguarding National Security Ordinance," March 19, 2024b.

Granahan, Colleen M., "Note: The Prospects of a Successful Reunification of Hong Kong and China in Light of the Events at Tiananmen Square," *Boston University International Law Journal*, Vol. 10, 1992.

Hakkinen, Teemu, "Parliament and the Gulf War in 1991," *History of Parliament* blog, January 20, 2016. As of February 12, 2024:
https://historyofparliament.com/2016/01/20/parliament-and-the-gulf-war-in-1991/

Halliday, Fred, "The Gulf War and Its Aftermath: First Reflections," *International Affairs*, Vol. 67, No. 2, 1991.

Helfont, Samuel, "The Gulf War's Afterlife: Dilemmas, Missed Opportunities, and the Post–Cold War Order Undone," *Texas National Security Review*, Vol. 4, No. 2, Spring 2021.

Helms, Jesse, "We Will Take Up Hong Kong's Fight," *Wall Street Journal*, June 20, 1997.

Holley, David, "China to Join Iraq Arms Embargo," *Los Angeles Times*, August 6, 1990.

Hong Kong Business Advisory, "Risks and Considerations for Businesses Operating in Hong Kong," U.S. Department of the Treasury, July 16, 2021.

Hong Kong Transition Project, "The First Five Years: Floundering Government, Foundering Democracy?" Hong Kong Baptist University, Government and International Studies, May 2002.

"Hong Kong's New Security Law Comes into Force amid Human Rights Concerns," Al Jazeera, March 23, 2024.

Hook, Brian, "British Views of the Legacy of the Colonial Administration of Hong Kong: A Preliminary Assessment," *China Quarterly*, Vol. 151, September 1997.

Huang Hua, letter dated 8 March 1972 from the permanent representative of China to the United Nations addressed to the Chairman of the Special Committee, United Nations document A/AC.109/396, March 9, 1972.

Human Rights Watch, "Hong Kong: New Security Law Full-Scale Assault on Rights," webpage, March 19, 2024. As of February 12, 2024:
https://www.hrw.org/news/2024/03/19/hong-kong-new-security-law-full-scale-assault-rights

Hurd, Ian, *After Anarchy: Legitimacy and Power in the United Nations Security Council*, Princeton University Press, 2008.

Ikenberry, G. John, *After Victory: Institutions, Strategic Restraint, and the Rebuilding of Order After Major Wars*, Princeton University Press, 2019.

Joffe, George, "Middle Eastern Views of the Gulf Conflict and Its Aftermath," *Review of International Studies*, Vol. 19, No. 2, April 1993.

Joint Declaration of the Government of the United Kingdom of Great Britain and Northern Ireland and the Government of the People's Republic of China on the Question of Hong Kong, signed at Beijing, China, December 19, 1984.

Jones, Clive, "Saudi Arabia After the Gulf War: The Internal-External Security Dilemma," *International Relations*, Vol. 12, No. 6, December 1995.

Kawase, Kenji, "Hong Kong Passes Article 23 Security Law with Sweeping Powers," Nikkei Asia, March 19, 2024.

Keller, Bill, "The Iraqi Invasion; Moscow Joins U.S. in Criticizing Iraq," *New York Times*, August 4, 1990.

Kozelsky, Mara, "The Crimean War, 1853–56," *Kritika: Explorations in Russian and Eurasian History*, Vol. 13, No. 4, 2012.

Lambert, Andrew, "The Anglo-French Crimean War Coalition, 1854–1856," commentary, Royal United Services Institute for Defence and Security Studies, March 16, 2010.

Lambert, Andrew, *The Crimean War: British Grand Strategy Against Russia, 1853–56*, 2nd ed., Routledge, 2011.

Lambeth, Benjamin S., *Desert Storm and Its Meaning: The View from Moscow*, RAND Corporation, R-4164-AF, 1992. As of February 12, 2024:
https://www.rand.org/pubs/reports/R4164.html

Lau Siu-Kai and Kuan Hsin-Chi, "Public Attitudes Toward Political Authorities and Colonial Legitimacy in Hong Kong," *Journal of Commonwealth and Comparative Politics*, Vol. 33, No. 1, 1995.

Lemnitzer, Jan Martin, "'More Serious Than the Eastern Question Itself—the Crimean War Compromise," *Power, Law and the End of Privateering*, Palgrave Macmillan, 2014.

Lieber, Robert J., "Oil and Power After the Gulf War," *International Security*, Vol. 17, No. 1, 1992.

Long, Tom, *A Small State's Guide to Influence in World Politics*, Oxford University Press, 2022.

Machain, Carla Martinez, "Exporting Influence: US Military Training as Soft Power," *Journal of Conflict Resolution*, Vol. 65, Nos. 2–3, 2021.

Maddy-Weitzman, Bruce, *The Inter-Arab System and the Gulf War: Continuity and Change*, Vol. 1, Carter Center of Emory University, November 1991.

Maizland, Lindsay, and Clara Fong, "Hong Kong's Freedoms: What China Promised and How It's Cracking Down," webpage, Council on Foreign Relations, last updated March 19, 2024. As of February 12, 2024: https://www.cfr.org/backgrounder/hong-kong-freedoms-democracy-protests-china-crackdown

Malinowski, Pierre, and Elena V. Linkova, "The Crimean War of 1853–1856 in the Memory Space of Russia and France," *RUDN Journal of Russian History*, Vol. 20, No. 2, 2021.

Margaret Thatcher Foundation, "Saddam Invades Kuwait (24 Jul–2 Aug 1990)," webpage, undated. As of June 6, 2024:
https://www.margaretthatcher.org/archive/us-bush

Maxwell, Neville, "Sino-British Confrontation over Hong Kong," *Economic and Political Weekly*, Vol. 30, No. 23, 1995.

May, Sophie, "Ceaușescu and the Soviet Invasion of Czechoslovakia: The Early Years of Dealing with a Dictator," webpage, Association for Diplomatic Studies and Training, undated. As of February 12, 2024: https://adst.org/2021/01/ceausescu-and-the-soviet-invasion-of-czechoslovakia-the-early-years-of-dealing-with-a-dictator/

McConnell, Mitch, "Hong Kong and the Future of China Three Years After the Tiananmen Square Massacre," Heritage Foundation, June 5, 1992.

"Memorandum of Conversation: Secretary's Delegation to the Twenty-Third Session of the United Nations General Assembly," *Foreign Relations of the United States, 1964–1968*: Vol. 17, *Eastern Europe*, October 1, 1968.

Muir, Richard, "The Iraq–Kuwait Border Dispute: Still a Factor for Instability?" *Asian Affairs*, Vol. 35, No. 2, 2004.

Mukherjee, Rohan, *Ascending Order: Rising Powers and the Politics of Status in International Institutions*, Cambridge University Press, 2022.

Murphy, Cait, "Hong Kong: A Culture of Emigration," *Atlantic Monthly*, Vol. 267, No. 4, April 1991.

Office of the Historian, Foreign Service Institute, "Soviet Invasion of Czechoslovakia, 1968," webpage, U.S. Department of State, undated-a. As of January 15, 2023:
https://history.state.gov/milestones/1961-1968/soviet-invasion-czechoslavkia

Office of the Historian, Foreign Service Institute, "The Kellogg-Briand Pact, 1928," webpage, U.S. Department of State, undated-b. As of January 15, 2023:
https://history.state.gov/milestones/1921-1936/kellogg

People's Republic of China, *The Taiwan Question and China's Reunification in the New Era*, Taiwan Affairs Office of the State Council and State Council Information Office, August 2022.

Permanent Representative of China, letter to the United Nations addressed to the Chairman of the Special Committee on Decolonization, Annex 1, United Nations document A/AC.109/396, March 8, 1972.

Pollack, Josh, "Saudi Arabia and the United States, 1931–2002," *Middle East Review of International Affairs*, Vol. 6, No. 3, September 2002.

Public Law 102-383, United States–Hong Kong Policy Act of 1992, October 5, 1992.

Public Law 116-149, Hong Kong Autonomy Act, July 14, 2020.

Quality of Government Institute, "Data Source: Alliance Treaty Obligations and Provisions Project (ATOP)," webpage, Quality of Government Data Finder, undated. As of March 10, 2025: https://datafinder.qog.gu.se/datasource/atop

Ramzy, Austin, and Mike Ives, "Hong Kong Protests, One Year Later," *New York Times*, last updated June 22, 2021.

Rath, Andrew, *The Global Dimensions of Britain and France's Crimean War Naval Campaigns Against Russia, 1854–1856*, thesis, McGill University, 2012.

"Report to the British Cabinet Summing Up the Situation in Czechoslovakia," The National Archives (United Kingdom), CAB 128/43, August 1968.

Rickli, Jean-Marc, and Khalid Almezaini, "Theories of Small States' Foreign and Security Policies and the Gulf States," in Khalid Almezaini and Jean-Marc Rickli, eds., *The Small Gulf States: Foreign and Security Policies Before and After the Arab Spring*, Routledge, 2016.

Riddell-Dixon, Elizabeth, "The United Nations After the Gulf War," *International Journal*, Vol. 49, No. 2, 1994.

Rodman, Peter W., "Middle East Diplomacy After the Gulf War," *Foreign Affairs*, Vol. 70, No. 2, Spring 1991.

Rolfe, Heather, and Thomas Benson, *From HK to UK: Hong Kongers and Their New Lives in Britain*, British Future, Department for Levelling Up, Housing, and Communities, November 2023.

Rovner, Joshua, "Delusion of Defeat: The United States and Iraq, 1990–1998," *Journal of Strategic Studies*, Vol. 37, No. 4, 2014.

Rubinstein, Alvin Z., "Moscow and the Gulf War: Decisions and Consequences," *International Journal*, Vol. 49, No. 2, 1994.

Ryan, Curtis, "Between Iraq and a Hard Place: Jordanian-Iraqi Relations," *Middle East Report*, Vol. 215, Summer 2000.

Sadowski, Yahya, "Arab Economics After the Gulf War," *Middle East Research and Information Project*, Vol. 170, May–June 1991.

Schumacher, Leslie Rogne, "The Eastern Question as a Europe Question: Viewing the Ascent of 'Europe' Through the Lens of Ottoman Decline," *Journal of European Studies*, Vol. 44, No. 1, 2014.

Sciolino, Elaine, and Michael R. Gordon, "Confrontation in the Gulf; U.S. Gave Iraq Little Reason Not to Mount Kuwait Assault," *New York Times*, September 23, 1990.

Šebok, Filip, "Ideology and Geopolitics: Chinese Reaction to the 1968 Invasion of Czechoslovakia," Central European Institute of Asian Studies, August 21, 2018.

Shachtman, Tom, "It's Time to Abandon 'Munich,'" *Foreign Policy*, September 29, 2013.

Silk, Leonard, "Economic Scene; The Broad Impact of the Gulf War," *New York Times*, August 16, 1991.

Stoneman, Anna J., "Socialism with a Human Face: The Leadership and Legacy of the Prague Spring," *History Teacher*, Vol. 49, No. 1, November 2015.

Swaidan, Ziad, and Mihai Nica, "The 1991 Gulf War and Jordan's Economy," *Middle East Review of International Affairs*, Vol. 6, No. 2, June 2002.

Sweetman, John, *The Crimean War*, Routledge, 2001.

Temperley, Harold, "The Treaty of Paris of 1856 and Its Execution," *Journal of Modern History*, Vol. 4, No. 3, December 1932.

Treaty Between the United States of America, Belgium, the British Empire, China, France, Italy, Japan, the Netherlands, and Portugal, signed at Washington, D.C., February 6, 1922.

Treaty Between the United States of America, the British Empire, France, and Japan, signed at Washington, D.C., December 13, 1921.

Treaty Between the United States of America, the British Empire, France, Italy, and Japan, signed at Washington, D.C., February 6, 1922.

Tsygankov, Andrei P., "The Crimean War, 1853–1856," in *Russia and the West from Alexander to Putin: Honor in International Relations*, Cambridge University Press, 2012.

UN General Assembly—*See* United Nations General Assembly.

UN General Assembly Resolution—*See* United Nations General Assembly Resolution.

United Nations General Assembly, "Twenty-Third Session Official Records," 1,679th Plenary Meeting, at 10:30 a.m., New York, October 3, 1968.

United Nations General Assembly Resolution 1514 (XV), Declaration on the Granting of Independence to Colonial Countries and Peoples, adopted on December 14, 1960.

United Nations Security Council Resolution 660, adopted by the Security Council at its 2,932nd meeting, August 2, 1990.

United Nations Security Council Resolution 678, adopted by the Security Council at its 2,963rd meeting, November 29, 1990.

"Unofficial Enunciation of the 'Brezhnev Doctrine,'" in Jaromír Navrátil, Antonín Benčík, Václav Kural, Marie Michálková, and Jitka Vondrová, eds., *The Prague Spring 1968: A National Security Archive Documents Reader*, Central European University Press, [September 26, 1968] 2006.

U.S. Central Intelligence Agency, "The Invasion of Czechoslovakia: Impact on Yugoslavia," Office of National Estimates, October 9, 1968.

U.S. Central Intelligence Agency, "'Czechoslovakia: The Dubcek Pause,' U.S. Intelligence Assessment of the Crisis," in Jaromír Navrátil, Antonín Benčík, Václav Kural, Marie Michálková, and Jitka Vondrová, eds., *The Prague Spring, 1968: A National Security Archive Documents Reader*, Central European University Press, [June 1968] 2006.

U.S. Central Intelligence Agency, "The CIA and Strategic Warning: The 1968 Soviet-Led Invasion of Czechoslovakia: An Overview," September 1, 2009.

Valenta, Jiri, "The Bureaucratic Politics Paradigm and the Soviet Invasion of Czechoslovakia," *Political Science Quarterly*, Vol. 94, No. 1, Spring 1979.

Van Evera, Stephen, "American Intervention in the Third World: Less Would Be Better," *Security Studies*, Vol. 1, No. 1, Autumn 1991.

Wang, Maya, "Hong Kong Repression's True Cost: Ominous 'National Security' Budget Increase Spotlights Growing Abuse," Human Rights Watch, February 24, 2021.

Ward, Steven, *Status and the Challenge of Rising Powers*, Cambridge University Press, 2017.

White, Robin M., "Hong Kong, Nationality and the British Empire: Historical Doubts and Confusions on the Status of the Inhabitants," *Hong Kong Law Journal*, Vol. 19, 1989.

Wong, Fanny, "Battle to Believe in the PWC," *South China Morning Post*, October 19, 1994.

Wood, Pia Christina, "François Mitterrand and the Persian Gulf War: The Search for Influence," *French Politics and Society*, Vol. 10, No. 3, Summer 1992.

Yahuda, Michael, "Hong Kong's Future: Sino-British Negotiations, Perceptions, Organization and Political Culture," *International Affairs*, Vol. 69, No. 2, April 1993.

Yellen, Jeremy A., *The Greater East Asia Co-Prosperity Sphere: When Total Empire Met Total War*, Cornell University Press, 2019.

Young, Louise, *Japan's Total Empire: Manchuria and the Culture of Wartime Imperialism*, University of California Press, 1998.

www.ingramcontent.com/pod-product-compliance
Ingram Content Group UK Ltd.
Pitfield, Milton Keynes, MK11 3LW, UK
UKHW050851130825
7370UKWH00026B/489